Technology and Purpose

Technology and Purpose

✦

Data Systems and the Advancement of the Not-for-Profit Organization

Joseph R. Liberto

Writers Club Press

New York Lincoln Shanghai

Technology and Purpose
Data Systems and the Advancement of the Not-for-Profit Organization

Writers Club Press
an imprint of iUniverse, Inc.

For information address:
iUniverse, Inc.
2021 Pine Lake Road, Suite 100
Lincoln, NE 68512
www.iuniverse.com

ISBN: 0-595-26596-0

Printed in the United States of America

First and foremost, this book is dedicated

To Mandy,
Without whom there would be no reasons.

This book is dedicated also

To Ed,
Who taught me that anything worth doing is worth doing well.

To Bill,
Who showed me that there is a great story inside each of us waiting to be told.

And finally,

To all of the dedicated and hardworking individuals of

St. Joseph Parish,
Cockeysville, Maryland

Contents

FALL: PRODUCTION

An Irish Blessing

May the road rise up to meet you.
May the wind be always at your back,
the sun shine warm upon your face,
the rain fall soft upon your fields,
and until we meet again may God
hold you in the hollow of His hand

INTRODUCTION

Where I live, even buying groceries from the supermarket has become a "high-tech" activity. Good, bad, or indifferent, dealing with technology is no longer optional. Whether you're looking up email addresses, or looking up the UPC code for bananas, you and your organization will need to interact with technology. Are you a "tech person"? Probably you aren't. I am not by any means a programmer or a wire guy. You may not be, either. All the same, we each bring unique qualities to the table and are affected by technology every day. We interact with technology in the course of doing our jobs, communicating with others, and participating in a host of normal, non-technical activities.

It seems that everywhere you go today, they are selling computers. Office supply stores, consumer electronics chains, even department stores and TV-shopping networks are selling computers.

The issues and decisions surrounding a business computer installation are different from those surrounding a decision to buy a PC for your home. You may be surprised to read this, because many of us remember when things weren't this way. It used to be simple. You went to the computer dealer, parted with several thousand dollars of your hard-earned money, and walked out the door with a bunch of cardboard boxes. Unless the computer broke down, you probably didn't venture back into the store until you needed a new one. Neither the boxes nor their contents were concerned if they were going to be used to balance your checkbook, organize recipes, do the office accounting, or play chess. They just did their jobs, eventually became too old or too slow, and were upgraded or replaced.

Sometime during the 1990s everyone decided that it was no longer sufficient for the computer to be just sitting there alone on the desk in the back. The PC became a business communications device, and things have never been the same.

I have been designing and installing technology solutions for individuals, small businesses and non-profit organizations for almost a decade now. After looking at my position as a technology consultant, and now a technology manager for a Catholic Parish, I realized that there was a great need that I could fulfill. People in these positions need a guide-map for technology in organizations for which technology is not in the line-of-business. I decided to write this book for both of us, as we are ordinary people who maintain or coordinate technology in small-or medium sized ventures. You don't have to be a wizard to read this book, or to consider the advice it provides. While parts of this book deal more specifically to the needs of the not-for-profit organization, many of the planning and technology concepts are universal to any small venture.

Welcome

This book is concerned with small computer networks. Small networks are systems that involve somewhere between ten and one hundred machines, with an emphasis on networks with closer to the former number than the latter. These numbers are not fixed in stone. However, when a network exceeds roughly 100 computers, certain aspects of managing the system require more involved practices. It is generally understood that a network of a size nearing a hundred machines would require a full time professional administrator. Professional network administrators should already be versed with the issues described in this book. There is a large difference, though, between the one or two machine network, and the hundred-plus machine network. There are also a great deal of small businesses, and non-profit organizations that fall somewhere in between.

You may think that the issues surrounding a small-scale installation of technology are too small to consider, but small-scale installations have many potential problems and pitfalls. Some of these problems are shared with larger systems, while others are unique. There is a greater tendency in a small operation to buy "off the shelf" consumer-oriented products, and to buy exclusively, or predominantly from consumer-oriented channels. A mix of inadequate planning and needs assessment, ill-developed policies, and even technical misunderstandings can cause the needless waste of time and money. More important than time and money, though, is the usability and enjoyment of the technology by the users. The biggest misconception is that *"Technology is cheap, and changes rapidly. We'll have to trash this stuff in two years anyway, so it doesn't matter"*.

Well, it does matter. Decisions made today will affect the future of your organization. Many disasters can be avoided entirely by asking enough questions, and listening to the needs of your people, and your organization as a whole.

Some of the landmarks on this roadmap are going to be familiar. If your organization has planned any type of capital project, such as a new building addition, or a major overhaul, you may have seen some of this information before. Many of the concepts in this book are based on content from Computer Science and Management curriculum, which is no surprise because this is my background. Although taught to computer science student, these concepts can be great tools for anyone to use as they plan a system.

The good system manager is not the one who never makes a mistake, but is rather the one who can regroup calmly and swiftly as soon as a shortcoming or problem in a system becomes apparent. Keep in mind, that much of this book is based on my experiences. I give you no "rules" for system administration. It is my intent to chisel down some of the key stumbling blocks found when adopting technology. Whether technology is in the form of computers, other office automation products, or telecommunications equipment, there are goals and plans that should be addressed.

Writing this book has been a great aid to the solidification of concepts and practices, some of which I have not thought about in some time. It has been fun, because every job has stories of success and failure, and these stories should be shared. Some say that hindsight is 20/20, and I only wish that I had had the benefit of experience when I began this journey.

This book is designed to aid you in creating needs analysis documents and understanding your requirements. It is an aid in defining specifications, Request For Quotes, and to help guide you through the initial stages of a networked system. In addition, later chapters handle the transitional aspects of implementing a new system, discussing issues such as policy development and training.

Wherever possible, I have tried to avoid overly technical explanations. While some technical explanation is necessary to illustrate key points, it is not my intent to create a technical manual. I want to focus rather on important considerations with business and management people in mind.

SPRING:
CONCEPTION

○ ○

"So little done—so much to do"

—*The dying words of Cecil Rhodes*

Ten people sit in a small room under the buzz of fluorescent lights. They look over stacks of proposals. They shake their heads. They shout their ideas, making their points clear. They argue. They laugh. Afternoon transforms into night. They go home, knowing that there is still work to be done.

These are the visionaries. During the day, one of these people is a CFO. There is also a software developer. There is the entrepreneur, speculating in the infantile age of wireless technologies. It is the heady days of the late Nineteen-Nineties.

These people have been given a mammoth charge. They are a committee. There must be a feeling of absolute power and potential in this room. Here, they will define a new system, one that will soon become pervasive, carrying this place through next twenty years.

A BRIEF HISTORY OF DATA-SYSTEMS

The term "Computer" was not intended for electronic machinery, but rather as a job title. In the dark days before pocket calculators, A computer was one whose position in life was to work, most likely in as part of a team of "computers", to do portions of complicated mathematical equations. In this way, high-level math people were able to supervise these teams of computers. Thus, these mathematicians would be able to focus on the more esoteric issues surrounding equations. Esoteric ideas that might otherwise be lost as these equations would take a lifetime of man-years to complete.

Electronic computers were not originally conceived as systems for data manipulation or intercommunication. The boon of technological development was primarily as a tool for military and scientific use. Much in the same way as the old teams of "computers", and preformed calculation tables were an aid to a mathematician, the new machines would lighten the workload for a new generation of mathematician. In the beginning, electronic computers were designed in the realm of the scientist and mathematician, primarily for the purpose of performing rote mathematics.

During, and after the Second World War, the realization of the potential power of electronic technology was realized. Rather than to rely on people and complex tables for calculations, machines were being developed to handle this work. Universities and corporations were developing systems to do basic math. A nation soon realized the multitude of applications of their technology, in the military arena.

Meanwhile, in postwar America, a different kind of problem was beginning to surface. Our victories in the war brought a return home of many young people, ready to start a new life. More money was being spent, and many companies flourished. Businesses were expanding into regional and national operations, and thus a more customers were involved. The amount of paperwork and transactions was increasing exponentially. To manage information electronically became a necessity as time progressed. Especially in the finance and banking industry, where more controls had been implemented, the issue of making data management and accounting work was becoming a serious problem. At the time, business' problem wasn't so much qualitative analysis of data, so much as the manpower required to achieve systems of reliable storage and timely retrieval of information.

The unlikely marriage of the scientific community's tools to the problems of business took a great deal of time to happen. Large electronics and business-machinery companies such as IBM, RCA, and others, as well as many others in a way defined this industry for many years. Usually, the door was the one commonality between business and science, mathematics. While electronic data processing has been an integral part of commerce for almost half a century, its roots are in the accounting department. With all of the mind-numbing accounting that a large company must engage in, the need for mechanization of some of these functions was obvious. IBM, in fact, was marketing computers in the fifties as nothing more than "Accounting machines". These, at the time, powerful systems were built to do nothing other than payables, receivables, and payroll processing—The basis of business management.

It took decades to shake this perception. During the nineteen-seventies, many types of companies, especially in the retail industry, were capitalizing on the power of electronic data processing. While the accounting department controlled most of the computer mind share, other departments such as inventory, control and other managerial functions were beginning to jump on board. A department store could, relatively quickly, find out how many avocado green leisure suits, size 40 Regular, were still on the rack at the Hoboken, New Jersey store. Firms were even beginning to tie inventory management systems into the Point Of Sale (POS) systems at each store. At the end of each business day, large reels of magnetic tape would be sent by some means, first physically, later in electronic form, from the mainframe computer at the retail store to the home office. During the night, the home office computer operators would tabulate all of the sales from each store, determining what needed to be ordered for each outlet. As time progressed, it became easier for management people to analyze the data. A Buyer for a national retailer could look at sales, and see that more red shirts sold in their eastern region than in the rest of the country. Thus, they could use this information to provide a greater selection or more inventory at this region, if they so chose.

You will notice that many of these examples involve large monolithic corporations. Nowhere mentioned is "John's Produce" or "The Corner Video Store". The barrier for data processing in this era was clearly the cost. In order to run a system of this size, one had to furnish a team of people, as well as an account with a Honeywell or IBM to furnish the hundreds of thousands, or millions of dollars worth of computer equipment necessary to set up a computer system. There simply was no middle ground, you either had the Big Dog, or you had no dog at all.

Thanks to research organizations like those at Bell System, This all began to change. The electronics industry was moving towards solid-state transistor-based electronics. Prior to transistors, the only reliable means of building an electronic computer system involved large numbers of vacuum tubes. Vacuum tubes were the state of the art, and allowed truly digital computing to be possible, but they had significant problems. They were relatively large, and required large numbers of tubes to create a useable piece of equipment. To call these machinations reliable requires some literary license, as statistically tubes were not really reliable, requiring replacement. Ten or fifteen hours of workable time between system failure was generally considered reliable. Because of the high levels of required maintenance, special rooms and teams of people were established to house and maintain computer systems. People requiring "data processing" services sent their data and programs to these data processing centers, where it would be processed by the operators of the computers, and eventually returned to the requestor.

For the business community, this was seen as "management" of the system. The scientific community saw it as this form of technology's Achilles heel. On one hand, the systems could perform calculations much faster than any person, but using their services required a host of additional steps that robbed the user of any performance gain. There was a noticeable void of "real-time" interactivity.

A computer company called Data General realized this void, and created a computer called the Nova. By all practical purposes, the Nova was one of the first truly "personal" computers, as it was small enough to be portable, and inexpensive enough to relatively affordable. There was only one problem, though. The Nova was really designed more for the scientific community. While this was a great advancement for colleges and science, it did not have much of a direct impact on the activities of business.

In 1964, a start-up out of Boston built a small computer, the PDP-8. This company was called Digital Equipment Corporation, or DEC. Digital basically invented, for more or for less, the first "personal" line of computers. They were called, aptly, the "PDP" (Personal Data Processor) series. Digital coined the phrase "minicomputer", and changed the way people and organizations viewed technology. Digital's computer was radically different from anything that other [mainframe] computer companies were producing at the time, not because of reduced size, but because of a fundamental change in the paradigm. Digital's computer was different because while it was designed for use by a single individual, rather than by a group of people, it was scalable as either a single-user computer or a group system.

Prior to Digital's PDP computer, most computing was still designed for a "host computing" model. Host, also called time-sharing, designs dominated computing for almost three decades. Time-sharing implies a scheme that is rarely seen in new applications, although is making a comeback in the form of the Web. However, in the 60s and 70s, it was borne more out of necessity than convenience. Time-sharing was a macroeconomics model if there ever was one. It was really borne out of supply and demand curve considerations, and not because people like to share or wanted to communicate. The connection between time sharing and inter-user communication was purely incidental.

Computer Time-sharing is based on the underlying reality that computing time and hardware resources (Cycles of processor time) are scarcer than the users who wish to use them. Until computers became ubiquitous, computer time was a valuable asset, and not to be wasted. Depending on the speed of the equipment, there are a varying number of cycles per second. Now, we call these cycles "megahertz" or "giga-hertz". In those days, a one Megahertz processor was something to get exited about. Today, typical speeds have exceeded one gigahertz, and are exponentially faster than computers of that era. In the early days of computing, technologies having to do with processor speed were in their infancy, and the microprocessor which helped speed and econ-omy along greatly, while invented, had yet to be widely used in a per-sonal computer application.

There were many time-sharing schemes, but they all shared one commonality: Many People using slices of the computer's time. The computer would go round robin, taking a brief amount of time (microseconds) for each user.

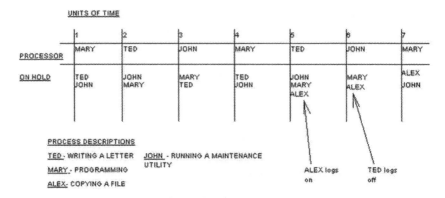

FIGURE 1.1: TIME SHARING

Time-sharing is kind of an interesting arrangement, because computers (even of the ancient variety) operate much faster than we can interact with them. Internally, A computer is very fast, and can do many things between each of our keystrokes. The premise of time-sharing was that the computer might as well serve multiple masters instead of wasting its precious time waiting for the (relatively) "slow" users.

One aspect of computers at this time was that teams of people were required to run a computer system. Additionally, large amounts of electricity and physical resources, such as space and air conditioning, were required regardless of how much use was made of the computer. Time-sharing allowed many uses of this extra time. Sometimes, companies would try to offset their cost of the equipment by renting use of their systems to other companies, or donating it to local schools and colleges. If you look at this histories of most all of the genius of the Microcomputer age, you will find its roots in the time-sharing system. Time-sharing also allowed accounting of "time" that a something was being done with the computer. Just as individuals are billed for electricity use or long-distance service, internal departments of a company would be billed for their slice of the overhead of running the company's computer. Accounting for this bill would be handled differently, depending on the nature of the company. If a firm ran its own computer, billing might be a budgeted internal overhead cost. If the computer was owned and operated by a third party, or contracted, the bill meant that real money needed be spent.

As long as the paradigm for system engineering was based on a terminal/host computer, the network will be a centralized model, with a group of people running the computer from one location, and controlling its use to one degree or another.

From a support perspective, this was not bad. The computer people knew where the computer was located, and detecting problems were simpler this way. The real problem was that the preciousness of the time precluded much tinkering with the technology. When time was costing money, the computer systems were treated as a resource that should not be wasted. Not surprisingly, access to computer equipment was not widespread, and development of new uses for technology was generally slim.

THE PERSONAL COMPUTER

Xerox was really the first company to successfully develop a computer system that truly moved away from the host/mainframe model, while still maintaining connectivity and the "network" model.

Xerox, in the early seventies, provided funding for a subsidiary research division, called PARC. PARC stands for Palo Alto Research Center. Although at the time, a large company doing basic research was not abnormal, PARC had some of the great technological minds of the day. Many established technological-industrial firms had been setting up facilities for performing research, especially firms claiming some kind of link to national defense, during and after the Second World War. PARC was not designed as a defense product research lab, but as a place to do commercial office research for its own sake. Bell System had Bell Labs, doing ostensibly the best research at the time in electronics and new telecommunications techniques. RCA had General Sarnoff's research labs, founded with the mission originally for war-related research. RCA, in time, developed amongst other things certain key aspects of the technology behind laser discs and, eventually compact-, and digital video disks. Like many of these labs, Xerox's PARC maintains a "college-campus" like atmosphere, in Palo Alto, California.

Possibly for the first time, a large entity was looking at computers as being communications and productivity tools that one person could use, for purposes other than the mundane tasks of math-oriented activity, like accounting, payroll, and inventory control. The people at PARC designed independent, connected workstations, for use by secretaries, managers, office workers, and a new breed of label, the "Knowledge worker". They had computers on the desks of the middle-management people, using electronic mail, typing letters themselves (a paradigm shift in and of itself), and many other things. What's the real kicker? All of this innovations employed systems that were remarkably like today's computers, complete with windowed operating systems, recognizable keyboards, and mice. Remember, this was the early seventies, most of this stuff would have seemed really Disney-landish to ordinary people. The most interaction most people had with computers in the seventies was the punched card that came with the electric bill (Please Return with Payment. Do not Fold, Spindle, or Mutilate).

This paradigm shift is an important cultural breakthrough. Up to this point, business as a whole was much more hierarchical in structure. The use of computers (or anything with a keyboard) was viewed as clearly a *clerical* function (note the negative undertones). Typing was not a desirable activity for one who was seeking a professional business career, or one who had his eye on the corporate ladder. Managers typically wanted no direct contact with computers. Until the proliferation of PCs, most did not even know how to go about using the computer system. It was conceivable, even likely, that anyone over the level of middle manager in a firm would have no direct contact at all with a computer system.

PARC's vision was the idea that technology was rapidly becoming more available, more viable, because daily it was becoming less expensive. Why was a company based in photocopier technology doing this kind of research? Xerox envisioned a day when paper would no longer the primary medium, and was trying to remain competitive. They wanted to drive the forefront of the new technology, applying resources to moving to a "paperless office" scheme. PARC's people designed a system whereby computers on each desk (real computers, not terminals) would be interconnected and could share each other's information and printers. This is the model involved E-mail, Graphical Interfaces, networks, and Laser Printers. Technologies that we take for granted today—25 years after they were first conceived in the Xerox labs.

Others were beginning to see the potential of technology as well, sometimes people within PARC. Bob Metcalf perhaps is one of the most famous of these people. Metcalf invented the means by which these "Alto's", the PARC personal computers, were interconnected. When he realized that the company was not going to do anything with this technology, Metcalf took his invention and founded a company called 3Com, incorporated. 3Com has been a leader in networking standards for computer hardware over the entire life of the personal computer.

By the late seventies, In parallel to the work being done at PARC, there were many companies making various marketable microcomputers, systems that were designed specifically for use by individuals. Xerox was not one of these companies. Due to the culture surrounding the companies building these machines, there was a backlash away from the mainframe/network model to the mentality of individualism and disconnectivity, a direction that lasted the better part of a decade. This was because of the idea that mainframe computers were not to be trusted, and that the individual should control the entire system. With independence from a network, or an Information Technology department, one could program or use the system as he or she saw fit. Connectivity was seen as undermining this element of individual control, and was not originally a popular idea. It really took a compelling reason to move back to this networked paradigm. This idea really came from small business.

Because of the nature of the companies selling microcomputers, there was a feeling amongst many business people that they were small, and thus worthless for business purposes. This was not completely unfounded. The early computers brought out were basically hand-built by very small companies. Usually neither of the two or three people involved in the business had any idea or desire to market to industry, and thus they were limited to the hobbyist market. Being limited to the hobbyist market was not such a bad thing, either. The state of the art in the late 70s precluded personal computers from becoming power-house data-processing systems. The big choice through the seventies was Cassette tape or floppy disk for storage. The systems did not have enough memory capacity or processor power to be good for medium- or large-scale operations. Because there were no standards in micro-computer systems, unlike the mainframe industry, there was no provision for interconnection. There was no communication whatsoever between software companies or between computer hardware companies. Until IBM stepped into the field in 1981 with their Personal Computer, no effort was made to develop inter-company standards. When the user community clamored for such standards as a common Bus (the slots by which add-on products could be connected) and operating systems, these requests were usually met with hostility.

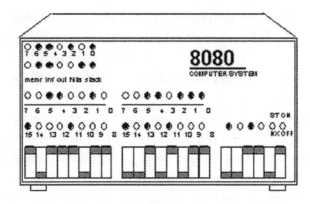

Figure 1.2: Early personal computers required that programming be performed by manipulating banks of switches on the front face-plates. The computer's response was indicated by a series of flashing lights. A cumbersome arrangement, this was quickly replaced aug-mented by add-on keyboards and video displays.

Although the intended market of such machines was the individual hobbyist, or home user, many personal computers were surreptitiously finding their way to business settings. The economics of scale were working, especially as the nineteen-eighties progressed and the IBM PC and DOS were legitimizing (and advertising) this technology, as the safe "standards" in the business arena. As more people entered the world of computing, the quality and range of software was increasing.

Very little of this revolution had to do with large companies, who were still tied to the mainframe model, Data Processing departments, and a universal mentality that the decentralization of the computer system would cause upheaval in an organization. This did not appear to be a revolution for an GE or a Boeing, it was a revolution for "John's produce market". John's Produce didn't have Data Processing, he had a hard enough time finding reliable kids to unpack the boxes of fruit. He also probably didn't have any bias against computers.

To do accounting, John could use an accounting program. John could type letter a demanding payment from a delinquent account, or use a communications program to order more peaches from the supplier. John could also keep a day-to-day inventory, if he chose, of his products. Using a minicomputer, John may have been able to do this before. The PC, however, enabled him to do these tasks (A) whenever he wanted to, and (B) without fear of leasing or time-sharing bills, and without the complications of a mechanism with a much larger organization in mind.

Surprisingly enough, John was not especially impressed by this new-found power. Millions of Johns, Toms and Joes were beginning to hear that there were great things going on. Things which could *potentially* be done with computers, but that there was very little guidance or structure. Buying a computer in the late seventies was a crap-shoot. Many systems were sold by funny companies that sounded like they were the names of planets from old episodes of "Star-Trek". Computers came disassembled, in kit form, so that the new owner could have the "fun and satisfaction" of putting the system together. Maybe even get a little high on the fumes from the solder. A great deal of Electronics knowledge and time was required to get these systems to run. Sometimes, computer kits didn't come with all the parts necessary. This purportedly added to the fun. ("Gotta make another trip to Radio-Shack for capacitors, Hun!"). Manuals were either non-existent or so convoluted that they might as well be. There were no spreadsheet programs, word processors, or databases. What was the most advanced and reliable mass-storage system? The ordinary table-top cassette recorder. Where was the progress? People who were willing to drop thousands of dollars for a kit from some no-name company, *and* could figure out how to put it together so that it worked, *and could* figure out how to program it had to see the need for good software.

Unlikely? But build it, they did. At the beginning of the 80s, there were hundreds of prospectors panning at this new stream. It seemed that, next to games, business productivity software stole most of the mind-share. Slowly, people were realizing that within "these silly electronic toys", there was a real utility, one that had never before been so easily accessible. And computers did indeed improve. Companies (some already established, some startups) were building complete computers which had software and accessories. For a few thousand dollars, places with one employee could keep track of their business to a degree that they had never known before. Professional people could correspond with each other, small business people could keep their books, and people like myself could write books. With a good enough printer, we could even produce them ourselves, if we so chose.

So, what happened? As a society of a million users, we decided to go back to the mainframe model. There was a void in the life of the 80's computer user. The PC was a greatly productive tool and a fabulously entertaining toy, but it was a really an island. Fairly early on, people were writing software, some of which would run an online community, and thus would allow people to use their home telephone lines to call in and communicate with each other. Scads of homegrown "Bulletin Board Systems, or BBS's" were popping up everywhere there was a couple of telephone lines, with the beginnings of online chat and electronic mail.

Once the computer started becoming pervasive, it was a natural progression to want to interconnect. People wanted to be able to use them for purposes other than writing letters or organizing their recipes.

So, here we stand, at the cusp of another revolution. Keeping score? (John's Produce: 1, Data Processing: 0). In many cases, the mainframe and data processing departments have either died, or have had to completely revamp themselves into a perverse type of handholding called "technical support". People are becoming used to, and intimately involved with, activities like video production and CAD that were out of reach only ten years ago. Businesses large and small, and organizations like yours have embraced this technology. You are probably thinking about doing so, and probably not "if", but "how" and "for how much".

But, How did it Happen This Way?

The Power of Thought

Return for a moment to the early eighties. Hey, return all the way to 1973, while you're at it. Xerox is clicking along, working on this graphical interface. People are buying into this machine as a computer for ordinary type person.

They had developed the first successful system with a completely graphical interface. To people other than computer geeks, this is terribly important. The GUI allowed you to get something referred to as WYSIWYG (What You See Is What You Get), Yeah, "WYSIWYG" is an actual technical term. What this meant was that Bold actually looks, well, **Bold**, *and*, ***Bold-Italic looks Bold-Italic.*** Bigger fonts got bigger on the screen, *and* on the paper printout. To what degree was WYSIWYG being used with literary license depended on the degree of precision required by the user, and on the system being used.

I know, I know, *all* computers do that! Well, in 1983—even in 1993 for that matter—all computers didn't do that. There were a whole host of non-graphical systems. DOS is probably the most famous. In many DOS programs, BOLD looks like*BOLD*, and Bold-Italic, well we're not even going to go there.

Apple Computer, in its early days, was able to negotiate an in-depth tour of Xerox PARC (where all of this magic was occurring). Apple's people liked what they saw, and decided to try to integrate the concepts of the GUI into their new product, called the "Lisa".

In 1983, Apple introduced the Lisa, for $9995, or "Just Under $10,000". As you might guess, Lisa was not a resounding success from a sales perspective. It was, however, a revolutionary product, nonetheless, from a technological standpoint.

Then, on January 24 of the following year, Apple introduced the Macintosh. Macintosh was a huge sales success, if not initially. The Mac changed irrevocably the course of history for computers and data management, and its manners and technology have driven computer systems ever since. Things would never be the same again. Microsoft, and a host of other companies, began introducing similar graphical operating system products for the PC platform. The rest, they say, is history

Apple computer was an organization with vision and a highly important product, technical as well as cultural. The Mac was not necessarily the fastest computer around at the time, nor was it the computer with the biggest hard disk drives or the most add-on equipment. There were other computers with better networking capabilities, even more robust operating systems. It was the interface that brought computing to you and me.

The power of the interface laid in how it *empowered* the ordinary person. There were no UNIX computers that empowered ordinary people. Likewise, there were no command-line based computers that empowered non-technical people. There were any number of other graphical systems, at the time. The counter-intuition of these systems tended to cause irritation and confusion, not empowerment. The Mac's interface was *designed* to be simple to use, and foremost to be a powerful system for the purposes of creation and communication. Macintosh computers were designed with people in mind, and the Mac did not get in the way of getting the job done. In many respects, these systems of the late 80's and early 90's were more elegant systems than what is available today.

It was in this spirit that people began to embrace technology. Not geeky tech types, but real people. Architects, Educators, even Bankers (of all people) began to use these machines. Graphic designers and artists found that they could draw on the computer, and the Mac became a canvas. Writers could create entire books. A kid with Lotus 1-2-3 could make charts and pie graphs impressive enough to take over the whole company, and composers could create music in a way which had never been envisioned before.

A company called "Aldus" appeared with a program called "PageMaker". Pagemaker allowed one person with a three thousand dollar computer and a five thousand dollar laser printer (remember, this *was* 1990) to do 80 or 90% of what previously required years of education, and an entire department of typesetters and printing presses. Aldus and the Mac brought the power of publishing to the people. The difference between the old regime and the new was that the computer had found a place, a compelling utility. Desktop publishing allowed the small-newspaper editor to create a level of quality in appearance, which was previously unknown to papers of this size. Small businesses and social organizations were able to do things and reach audiences that were previously untenable. The power of applied technology brought a level of empowerment and communication previously unknown to any of us.

So, How do I play into all of this?

Well, So how did I become a part of this? I was in a generation of kids who were the first to be saturated with technology. Like most kids, I discovered technology at school. I can't remember going to school where they didn't have a computer lab. Banks of beige Apple II+'s and IIe's, and those weird, black Bell and Howell Apple clones lined the walls of the computer labs in the school that I attended. Like millions of other kids, I learned my way around the Apple II with the likes of Carmen Sandiego, LOGO and it's amenable "Turtle", BASIC, and Print Shop. They even had a few Macs, but these were off to the side, and weren't really accessible.

I really didn't like the Apple II. Like many other kids', My parents had an Apple IIe. It sat in my dad's den, and I logged many hours with AppleWorks to type papers, and such. I could never really dig the keyboard, or the green monitor, but most of all, the II never let you work in any two things at once. Don't get me wrong, it was a good computer, but I never felt a connection to it.

In 1987, everything changed for me. The local community college had a computer camp in the summer for the kids in the area. They taught programming and computer logic, and other nerdy stuff. They had a Mac lab. The blue glow of a parcel of little Mac SE's, the understated symphony of their fans, and the hollow, irregular clicking noise of thirty mice permeated this room. This was high-tech, space age technology. There were no funny, floppy disks. These machines had hard disk drives, and good software. The room was surrounded by glass walls, and was in the center of a much larger lab of Apple II and IBM computers, which only added to its aura. I still remember the first Mac program that I ever used. It was called "Comic Strip Factory". Comic Strip Factory was a truly great program. You were presented with a blank layout board, and could choose from a host of truly brilliant art that you could use to build comic strips. You used the art to add the characters, background scenes, and put the little dialogue balloons for everyone. You created the cartoon. It was better than a game.

Well, that would the last Mac that I would use for three years, until the spring of 1989. I was in the eighth grade at St. Paul's School in Baltimore. The school said to the parents that everyone going into ninth grade would be advised to go and buy a Mac, because that's what the school had. My parents protested, because they "already had a computer". They bought it anyway. Sometime early in 1990 I had a brand new Mac LC, a 12" color monitor, and get this: A Laser Printer! This was the first new computer that my parents had bought in almost ten years. It was light years ahead of the IIe.

St. Paul's was somewhat progressive at the time. They had a few classrooms with no desks. People could sit around on the floor, or in the corners of the rooms. They had established the top floor of the building as an open computer lab, called the "Student Writing Center". The writing center had been set up a few years earlier, and was a huge room with thirty Mac Pluses. Still fairly recent machines, it was like scotch to an alcoholic. There was no regular supervision, and no sign-in sheets. These computers were open season. This was also the first experience that I had with a network. They had a powerful file server with something like 100 megabytes of hard disk space, located in the back room, which served as a place to store files, and to run programs.

They had a good arsenal of software. They had word processing, page layout programs, illustration programs and this graphics program called "Fullpaint". Hanging out there until well in the evening, I cut class regularly to the point of almost failing, I lived in this room. I learned how the file server worked, how to copy the programs off of the server, and made friends with the other regulars. Shufflepuck, a virtual air-hockey game, was my downfall. This would be my last year at St. Paul's.

Going into the tenth grade, I found myself in a different atmosphere, in the Catholic School System. Loyola High School in Baltimore had no Macs, except for an ancient 512k with no software in the library. A priest, a wizard who had been teaching technology for thirty years, and had forgotten more about programming than I would ever know, ran their labs. Fr. Murray would be an inspiration, because it was he who suggested that I revive their computer club. We would meet once a week for an hour in the afternoon. We could fix equipment, or just hang out. The club eventually got a donated computer and a little bit of funding for a phone line, and set up a bulletin-board-system. Students could call in and chat, or get information.

The Internet was becoming known about, and we had a connection link from our bulletin board system to an internet provider called "Sailor". Sailor was a free service that was part of the public library system.

College, and Contracting

As time progressed, I began to get more involved in technology, eventually playing a key role in the IT department at Loyola College, supporting the Macintosh systems for the college. I was also starting to do a great deal of contractual support for small businesses. Somewhere along the line, I even went back and worked for my high school, fixing the older computers that I had used as a student. Loyola College was different in that they had a real network. I owe a lot of my knowledge to my three years working there. I learned a lot about how networking computers can deeply enrich the user experience. I also learned a lot about how Information Technology departments work, and sometimes don't.

After college, and a brief stint in professional contracting, I found a job with a non-profit organization, a beautiful Catholic church and K-8 School. I have worked to set up an information systems department, and a reliable networked data system. The diversity of people and ideas in this environment is inspiring, as is the varied responsibility and challenge in a system of this size. I have looked at every step of my journey so far as a rung in a ladder. Each new experience adds new knowledge and experience to my life. There is a greater reward for me in working here, because the people here are working every day to make others' lives better, or sometimes worth living. There are very few corporations that can say as much.

This is the application of technology with a purpose. Not the purpose of making a few dollars, but the purpose of social action, education, and the improvement of community and society.

The Tao have a proverb; "the journey is the reward". I have tried to live my life with this in mind. This is but one step in a very long path of knowledge and experience. On your path, use the design of your system as a learning experience, a step along your path. Remember your experiences. Remember your mistakes as well as your successes. Even if you are not a technology enthusiast, there is a great deal of challenge and opportunity waiting for you.

As well as a great deal of reward.

THE FOUR SEASONS

Perhaps you have already danced with office automation. Maybe you installed a phone system last year. Maybe you have voice mail, computers, or a web site. It is likely that you or your staff have word processing equipment or computer systems on their desks. Maybe you use a computer already at home, and have decided to make the move at work. Perhaps you got an advertisement, or an enthusiastic suggestion left on your desk from one of your staffers (a circled $999 on the latest "Dell" ad).

"Why can't this be more straightforward?!", you think. It may be a realization of a cumbersome, time-consuming task or of an integration void between your existing systems. The need for automation in your office usually stems from some organizational dysfunction.

Because my experience has been with non-profit groups primarily, many of my examples and scenarios surround this type of organization, and its specific needs. My book deals with things like Technology committees, which you are unlikely to find in commerce or industry.

Does this mean that this book has nothing to offer you, because you are in a for-profit situation? Of course it does not. This book has something to offer you, too, because in this way small businesses have a lot in common with Churches, Schools and other non-profits. Although much of my experience is with education, I have tried to stay away from many of the school-specific issues here, because there are many good books out there for school applications.

To understand where I am coming from, you need to know a little bit about what I do. I am a network administrator for a network of around 100 machines. My network offers some extended services, but is fairly straightforward, and has been reliable. We offer things such as

Internet access, electronic mail, and data storage. My position is primarily a managerial one, with some technical aspects as well.

If you are in a small organization, and are reading this, it is likely that you are the one doing technical troubleshooting and support for your organization. You may be a professional in the technology industry who was hired to run the network. You may have talent in figuring out problems, but no formal training in technology topics. Somewhere along the line, you unwittingly became branded as the "Computer Guru" in your office. Either way, you have since known no peace.

Networks large and small can be beasts, but in a small organization they can be especially a boon or burden, depending on how they are managed. Network administration is a full-time job, requires attention to detail, lots of documentation, and a person who feels a commitment to this responsibility. Many things are under your control, and a good administrator will not abuse this power. Backing up data, managing users, and running a good network is key to the happiness and satisfaction that others have in your organization. A good network can help to bolster productivity, a poorly run network will cause bad moral and inefficiency.

Typically, in a small organization, the needs of the data system become larger than the manpower and resources available to support it. In large companies, an entire team is available to support different aspects of the data system. Development, Infrastructure, Network Administration, Telecommunications, Management, and Training are common subdivisions of an Information Services department. Additionally, In an educational environment, there is typically someone who manages and develops the audiovisual system. This field, Instructional Technology, is the application of technology in the classroom.

Network managers in small systems wear all of these hats. One day a network manager might be preparing the budget. The next day it is he or she may be on a ten-foot ladder, up in a dusty ceiling adding a new cable. Where The precepts of the job, to maintain and develop the system, always remain the same the activities change from day to day.

This adds excitement to the job, but also a unique terror that the carefully managed balance of responsibilities can topple at any time, creating a monstrous amount of work and stress.

Why do organizations like these need computerization and network managers in the first place? Electronic data management, and electronic communications can replace the need for expensive mailings, rooms full of filing cabinets, and hours of drudgery in one's job. Getting volunteer help is a chore in and of itself, and lots of your volunteers or secretaries doing mindless chores, such as stuffing envelopes for a mailing, may be able to be eliminated, or significantly reduced. Secretaries, too, are often overlooked. A parish secretary or a secretary in a non-profit organization is the hub of activity. They are office managers, front desk people, and the mechanism by which the entire organization remains organized and up-to-date. Busy-work, not people, should be technology's target of elimination. Trying to use technology to eliminate people is a very bad idea. In our organization, we use technology to augment people, and to improve the range and productivity of these people within our organization.

Electronic systems can be used to manage scheduling, correspondence, and manage information about our community base. In my organization, which keeps certain information about its parishioner base, this function is crucial to business-continuation. Database products have been developed to handle the unique needs of parish data management, and have helped enormously to take a huge workload off of the administrative staff. In addition, smaller databases of volunteers have popped up, helping all of us to stay in better contact with each other. Used with a sense of balance, technology can be a great leverage tool for any community of people, small or large.

THE NEED FOR MANAGEMENT

There are many reasons why having centralized management of your technology systems is necessary. Management can provide the structure and direction that your system needs to be an asset rather than a liability. Proportionate to their level of complexity, they will require increased support and service.

A data system is a puzzle of interconnected services. Backup, Infrastructure, Servers, Clients, Email/Internet, and other systems interlock to build a puzzle of a system. Each individual piece of this puzzle is both crucial to, and yet no more or less important than, the entire picture. Centralized management structures the thinking about these services as building blocks of a *system*, rather than as individuals.

Decentralization of management creates a great deal of redundancy and focus on each issue as an individual issue, rather than as part of a system. The puzzle pieces will not be able to come together under these circumstances.

<u>Individual's or Individual Department's Purchasing Priorities</u>
Cost/Bundling Options
Vendor Preference/Ease of Ability to Purchase
Shipping costs, if applicable
Convenience
Ease of Use
Post-Sale Support and Maintenance

<u>Individual's or Individual Department's Needs</u>
Software
Hardware
Printers and Support Supplies
Modems
Internet Service
Email
Backup

When trying to deal with these purchasing decisions, departments may all be making the same decisions, using numerous different vendors or suppliers. One office may purchase everything from printers to paperclips from the same supplier. Others will go to the local electronics store and sign up for a package deal, which may include Internet access.

What you end up with is a system that cannot be readily interconnected. Interconnection is a key to efficiency and productivity within your system. All decisions, which are made with regard to any implementation of a system, should be made with Interconnection in mind.

Centralized technology management can deal with many of these issues. A central parts vendor can be selected, as well as a supply house for printer ink, and such. The increased volume may even make a beneficial difference in cost, saving your organization money.

Advantages of Centralized Management

- The ability to set up purchasing relationships with one vendor

- The ability to establish leverage with a particular vendor through economy of scale

- The ability to manage resources and make decisions from a higher viewpoint of the entire organization, rather than just one department

- The ability to devise and implement policies and standards

- The ability to cut redundancy

- The ability to develop long term plans for the organization

The First Question

Why is Technology Important?

A quiet revolution is taking place. We have been moving from paper books, ledgers and typewriters to computerized data management programs. Many major functions of an organization are currently moving to entirely electronic systems.

Whether it is a decision support system for the accounting department, a database for addresses and phone numbers, or grade book or billing management software for a school secretary, we depend heavily on technology in day-to-day life to maintain up-to-date information on our operations.

Telecommunications. E-mail, Web, Database, Groupware.

These are buzzwords many of us are familiar with, in businesses and in the modern world. Why is this level of integration important in a non-profit organization?

In the business of social action or education, as in other lines of business there must be an underlying reason for technology's role in an organization. Here, the common thread is the facilitation of communication. Whether through educational support, for the use of social justice, or even the varied decision support systems on which your accounting department relies, the role of technology will evolve every year within your organization.

Look around your own organization. Many groups use hand-made display boards, but look at the technology involvement in their production. Every year, more groups are beginning to use video and graphic presentation regularly, in lieu of slides or overhead projectors. People are starting to generate their displays on computer equipment, and use computers to generate flyers and posters. As younger, more technologically comfortable people begin to take the reins of leadership, electronic integration will begin to flourish in many varied areas.

The Community Safety-Net

In the early days of the phone system, communities decided that it was important that there be someone available, on the other end, to call for "help". Up to this point, there was always some kind of an operator who could in turn call the police or local fire department to report an emergency. The codification and implementation of life-safety systems, and the implementation of the national 9-1-1 system transmuted the telephone system—lest in the eyes of the populous—from a convenience to a necessity. Today, as an extension of this, hordes of people are signing up for cellular telephone service for the very same reason—safety. The telephone system is one of society's safety nets. It allows us to call the police, or mom, or the kids, no matter where they happen to be. Your organization is, by nature, part of society's safety net. The services which you provide make you a part of the community. There are people out there who you cannot help because *they* cannot contact *you*. Contact and communication are an important benefit of technology. Quickly, the internet is becoming an important part of the safety net.

Personal crisis can take many forms and situations. Perhaps the emergency is a loss of a loved one or a job. In many urbanized areas, use of computers with Internet access is being provided free of charge in the public libraries. People are moving to the internet to find jobs, and to find resources when they are in need. If someone is looking on the internet for the services that your organization provides, will they be able to find you?

"Why does our organization need technology?". Before the technology plan, before we talk about wire and computer networks, and all of the fancy services that can be provided, this question must be answered. It is important to understand why your organization needs technology, because technology, no matter how you do it, is expensive, and time consuming.

This should not dissuade you from adopting technology in your organization. There is a place for technology. Sometimes, though, having good reasons for its adoption can make the goals and needs seem clearer.

Training as a Mission

Perhaps your place in the safety net is as a resource for people in need. Technology, aside from its communications facility, may be implemented as part of the line of business, such as for the purposes of training. The mission of many a non-profit organization is to train people. Usually training is a tool for the greater goal, which is an improvement in the quality of life of some group of people.

Your group may work to support the needs of the homeless in your community. Part and parcel to improving these people's situation is to provide training, perhaps training in technology, perhaps other types of basic training, like reading or mathematics, typing, or other basic skills. Electronic training can work for an organization, but not without structure. Planning the needs of the training program should be paramount in setting up a program.

Training Programs may include:

- Basic skills, such as Reading or Math

- Labs for teaching specialized skills

- Job Retention or Clerical Skills, such as:

  - Typing

  - Word Processing

  - Spread Sheets

As with any education, there are priorities surrounding teaching with technology. It is crucial that your training facilities be as up-to-date, and most importantly, as consistent as possible.

Consistency is more important to both your instructors and students than being state-of-the-art. A mish-mash of hardware and software will make teaching any skill ineffective. If your organization, like many, is dependent on donations to fulfill its mission, this may be difficult. Corporations and higher educational institutions should be approached, with a specific goal written to give them an idea about the scope of your project. More about this will be discussed in the section later in this book regarding donations.

Specialized labs may be donated or based on grants from an organization which is a parent benefactor of your program. These types of programs may include job training for specific jobs: Training for participation on an assembly line, specialized programming or data entry for a particular kind of job, operation of a telephone system, and other like skills. Some specific company, or industry group, would sponsor this type of training, and would provide the equipment and programmatic material necessary to instruct. Usually, organizations provide this training on their own, and may coordinate with your organization as a channel for new people. Due to the inherently proprietary nature of this type of program, it is highly unlikely that your organization would attempt to implement such training without patronage.

Organizational Advancement

Tandem to training (or perhaps exclusive of it) is the likelihood that management or communication advantages can be gained through computerization. Technology can bring organization and control to situations where they may currently be lacking. A systems approach can help to shape and organize methods and structure.

Electronic Systems can aid in the:

• Creation of a databases to manage your client or volunteer base

• Financial Accounting Process

• Production of Mailings, Brochures, or other Literature

• Management of Correspondence

• Communication of your goals to the internal people and to the outside world

Technology in the Future

It is important to realize that what may have been in the past may not hold true for the future. Society changes, and just as the telegraph has made way to new technologies, other facets of communication are transforming and will transform in the future.

Youth oriented programs are beginning to see this today. Generational differences as they relate to communications are becoming terribly important to organizations. Reaching children today requires that you have web pages, chat-rooms, and email. The Reading the "Church Bulletin" seems as unlikely and ludicrous to many of them as checking for new information on a website may be to other, more established members of the community.

While this is still a socioeconomic phenomenon today, education and culture in America is changing. As time goes on more and more of the population will be tuned into this channel. Being literate, computer or otherwise, is a reality which cannot be avoided.

In addition, you can use this socioeconomic aspect to your organization's advantage. While your client base may not be looking for you online, in the future benefactors and resources are very likely going to find out about your organization through these means. Having an Internet presence is, and to an increasing degree will be in the future, as important as having a phone or a street address. Organizations and individuals will have only two directions to choose from: Participation or invisibility.

TECHNOLOGY PLANS

Regardless of your organizations goals, one of the first steps to starting a data system project is to ask your organization is if it has considered a technology plan. Technology plans can be very simple, maybe even a statement of where you would like to see technology implemented.

Usually, it is a good idea to stagger the goals, with some being near-term, and others being longer term.

Try to resist making your plan technical. Use broad strokes, referring to services and goals for the system. Usually, you do not want to exceed five years in a plan such as this, as the state of the art keeps changing. Five years from now, your plan will likely seem dated. The plan should be printed, and somehow made available to every key person in your organization, from the top level to the administrative level. Give people a chance to read the plan, and comment on its goals, because this is the easiest time to rethink a strategy, or reword a goal. After you spend money on equipment, you are already in the game. You may have already spent some money, and may have turned out lucky doing this. Then again, you may feel unsatisfied with what has been purchased, and wish you had the money back.

The goal of technology plans is to create a seamless, integrated environment for the user. Computer service, like any good service, should be akin to dining in a good restaurant. The water and wine should always be at hand, while the service remains as invisible as possible. Empty glasses or soiled dishes on the table, even an obnoxious waiter can ruin an otherwise good meal. Obnoxious computer systems can ruin the experience for the users.

Computer systems and Home Entertainment systems are very similar. Let's say you bought a TV in the 80s. It was a good set when you bought it, but not cable-ready, so you needed to get some kind of means of receiving the signal. You decided to get a cable-ready VCR. You bought different brands, so the one remote control wouldn't operate both the TV and VCR. If you wanted to watch television, you turned on the set with one remote, and the VCR with the other. You could use the VCR remote to control channel selection or tape playback, while you had to revert for sound control to the television remote.

Some time in the 90's, you decided that you'd like to have better sound than the TV set could offer you. You bought a stereo, which came with another remote control. This remote had no less than 50 features, and con-

trolled a host of VCRs, Laserdisc players, and other periphery, but only within its own brand, which you do not own. Now, you have a remote for the TV (Power), The one for the VCR (Channel Selection), and the one for the Stereo (Volume Control).

Last year, you bit the bullet and "went digital" with DVD. If you're keeping count, we've got 4 remote controls now. (the television, your old, creaky VCR, the Stereo, and now the DVD player). Your TV had no more connections available, so you had to hook the DVD player up to the VCR, which was hooked up to the TV set. Now, when you want to watch a DVD, you need, at least one time, to use all four remote controls. As if to add insult to injury, All of these remotes overlap each other, controlling optional VCRs, TVs, Cable boxes of their respective brands and eras. All are essential for the operation of their own equipment, but none are 100% useable and none are compatible with each other. This is because your stereo, TV, VCR, and DVD player are of different makes, and of different generations of equipment. The side effect for you, the hapless owner of this system, is that in order to control any one aspect of this entertainment center, you have to use no less than three of these controls.

All too often this is how technology works. Many people tolerate overlap and gaps in service, simply because they do not have the tools at hand to correct the real problem. This usually manifests itself in frustration with the system as a whole. You may find this situation with voice mail. For a while, I had voice mail on my work line, voice mail on my cellular phone, on my pager, and an answering machine. With this much redundancy, the voice mail had become a weapon rather than a tool. If you needed to get in touch, you could leave a voice message on my pager. Unfortunately, the pager was all too often left at home or work, and the messages would go unchecked. With all of this voice mail, there affords a threefold opportunity for missing the message than if I had only one number and machine. If you need to reach me, this is the worst possible of situations. Not only have I forgotten to check my voice mail on my pager and thus missed your message, but you harbor a false sense of security thinking that I will certainly get the message. Too many people have this unwieldy situation in their daily lives, all too often because they don't investigate their options beforehand.

USING THE SYSTEM DEVELOPMENT LIFECYCLE FOR PROJECT MANAGEMENT

Think of the System Development Lifecycle as the Four Seasons of Life. Spring, summer, fall and winter.

The spring offers an opportunity for conception, the planting of seeds, the planting of ideas and a brimming anticipation of potential on the horizon.

The summer brings fruition to the potential. The real labor begins in the summer. Bringing the ethereal world of plans and planting, and ideas to the physical realm. Summer is potential realized.

With fall comes production. The fall is your application of what has been planted and cultivated. "You shall reap that which you have sown".

Winter brings not an end to the cycle, but rather rebirth. Retirement and demise face the legacy of the previous year. A hope for the future is a glimmer in the eye of the past. Winter brings the reality of death and rebirth.

Bear in mind that we are not referring by any means to a chronological year. This process may span many years. Our system, for instance, is dead center in the Fall, and is seven years into this process.

Whether you are a farmer or a computer guru, without good planning, good seeds, and good tools you will not find success. *Many organizations find that ill-planned technology decreases, rather than increases productivity amongst staff.*

Go back and read that last sentence again, because it is important. You will find that technology becomes something that everyone uses more or less for their own purposes and in different ways. The problem is that when you realize the problems which arise from lack of structure, you will need to take away something previously given to your people. Outlining guidelines and structure from the beginning is a much preferable option.

Through the remainder of this book, integrated into the topics will be the various elements of this "four seasons" model. A model called the System Development Lifecycle, or SDLC inspires the structure of this book. The SDLC is a staple tool of technology project managers everywhere. Webopedia (www.webopedia.com), an Internet-based encyclopedia, defines the SDLC as

> …The process of developing information systems through investigation, analysis, design, implementation and maintenance. SDLC is also known as information systems development or application development. SDLC is a systems approach to problem solving and is made up of several phases, each comprised of multiple steps:
>
> - The software concept—identifies and defines a need for the new system
>
> - A requirements analysis—analyzes the information needs of the end users.
>
> - The architectural design—creates a blueprint for the design with the necessary specifications for the hardware, software, people and data resources
>
> - Coding and debugging—creates and programs the final system
>
> - System testing—evaluates the system's actual functionality in relation to expected or intended functionality.

The SDLC has been used to manage many large-scale software and systems development projects, involving many people. A large-scale SDLC project may involve hundreds of end-users, and dozens of software product developers. In a large-scale project, there is a necessity to have structure at every level, even in the methodology used to manage the project itself. Although in a small-scale system, there may be only one or several "developers" of a system, this basic structure can be modified to suit a variety of diverse needs. Organizational methodologies are like personal goals and time management schemes. They tend to be as effective as those implementing them. You may not be a "organization methods" type of person, in which case methodologies may not impress you. Keep in mind, also, that while this is a good model, it is also not the only model.

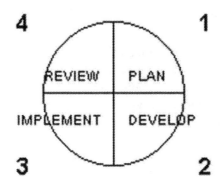

FIGURE 1.3: THE SYSTEM DEVELOPMENT LIFECYCLE

The System development lifecycle is a four-phase design. A project moves through the four numbered stages; Planning, Development, Implementation and Review. The fourth stage implies that the project implementation was successful. The review portion is important, because the project should never *end* at phase four. Technology projects in today's era are like children. There is always opportunity for both improvement and growth. Technology managers who install systems, and sit back and eat donuts until there is an obvious problem (such as obsolescence or a system failure), set themselves up for headaches and career problems.

For our purposes here, an expanded version of this cycle may be illustrative. It is my opinion that there are certain aspects which, while addressed in the original model, have not been illustrated adequately in the diagram. Issues such as establishment of an exit strategy in your IT projects early on, and documentation should be vocalized throughout the project. Here is an illustration:

FIGURE 1.4: THE SYSTEM DEVELOPMENT LIFECYCLE, MODIFIED

Because the System development lifecycle is a tool traditionally used in software development projects, its basic form above is modeled to that purpose. It is a malleable form and because it is a good tool, it has been applied to all different kinds of technology projects. Even if only to guide you in your process, using this model can help to point to what may be over the horizon when putting together a data system.

Bear in mind that, although it is the one used in this book, this model is not the only management tool available. There is always room for change and improvement, and a system is always in need of modifications and additions to improve its efficiency.

PLANNING

The importance of Planning cannot be overstressed. It is a good idea to spend most of your time in any Information Technology based project in the planning phase. It takes a lot of time and effort, but it prevents a great deal of trouble down the line. Maybe it starts with asking questions of the key people (possibly yourself) who are responsible for the high level processes which you have decided to consider automating. I have compiled here some preliminary questions I ask myself, then others in my organization. Usually I try to start with higher-level decision-makers, them move on to other key players in the organization. As with everything in this book, depending on your circumstances, you may wish to add questions, or modify this plan to suit your specific needs.

Write these questions down for each interviewed person, and conduct an interview. Maintaining a quiet and professional atmosphere is very important, as this is the first "official" introduction of your plan. Initially, you may wish to review the key points of the project. Keep a binder of everyone's questions and comments, along with everyone's specific answers to your questions. You may want to videotape or record the interview if you find it helpful. Preferably, ask these questions to all of the people independently who are pushing for automation. You may be surprised at the variety of answers you receive to these questions.

Important to this process is to try to gauge others' thinking about automation, and not necessarily to get parity in all of the responses. All people are different, and tend to see their departments as centric to an organization. I have found that even high-level staff will focus exclusively on their own needs and concerns, without any consideration of others in the organization. Specialization is generally a good thing, and not necessarily bad for an organization. You as a systems integrator (Surprise: that's you!), need to keep everyone in mind when you design your systems strategy.

1* *What* process *it that we are trying to automate?*

What is a process? Well, a process could be phone calls made to your offices, correspondence, or if you are in a medical practice, sending in billing information for a patient to insurance companies for payment. For each separate application, each of these questions should be asked

2* *Is this process Codified?*

This question is deceptively simple. You may need to do some leg work to find the real documentation behind a process. Training documents, and policy and procedure documents can be helpful. This part involves getting background information about whatever it is that you are trying to automate. If your organization has a policy manual, it may have documents on how this process *should* work.

A necessary question to ask is if there are controls or policies to protect the integrity of the process from infiltration or theft. Especially in accounting systems, part of the process usually involves auditing and control. This may have an affect on your automation plan.

3* *Who is involved in the process*

Make a list of all of the people and their respective contact information who are involved in the process. It is very important later on that everyone gets asked about their opinions, and feels to have been made part of the upgrade project. Indifference and confusion too often occurs when people are not asked to participate. Even in the smallest organizations, non-involvement can lead to, minimally, barriers regarding buy-in to a new project. Sabotage or negligence can also be side effects.

People are the most important part of any technology project, always keep this in mind.

4* *Who manages the process?*

Listing an organizational tree may be helpful at this point. It is important to know who has been in charge, and will continue to be in charge of their process. It is very easy, even inadvertently, to use technology to leverage power in a small organization. This benefits no one. Your responsibility, inasmuch as you are the system administrator, is as a manager and facilitator of the means by which the process is done, and not necessarily the manager of the process itself.

5* *How does this process actually work?*

Ask these people (especially the mechanical people behind the process) how it works. Bear in mind that there is a tunnel vision in many people's minds, especially at this level. You are likely to get a lot of extraneous information and opinion, but listen to what people have to say about a process, especially about its shortcomings.

Note: Are the controls and protections on the data from question 2 actually being implemented in reality?

6* *What is the culture behind the process?*

Is this process managed by the "graphic artists" in your firm, or by the accounting department? You may find a different set of rules depending on the answer.

A good example of this is a situation that occurred recently in the our system. We are a Catholic parish and school, and have built a great deal of new infrastructure over the past three years. As a church, we have many functions running constantly at our facilities. We keep track of these functions on a large paper calendar in our office. Anyone at the senior level of management "The staff" (about 12 people in our parish) can add/change/delete functions from this calendar. Everyone on the staff has certain scheduling responsibilities. Annually, the entire staff meets, to resolve room-scheduling conflicts and to look over the calendar for the next year. Our calendar year is 12 months, so nothing can be scheduled out more than 12 months from the beginning of the calendar year. This process had been in place for many years, unchanged.

We decided a year and a half ago to try to automate the calendar process. This was because the process was considered to be excessively cumbersome for the administrative people for whom the responsibility of producing this calendar fell. When I arrived, the calendar was made manually with a photocopier to enlarge a picture of the calendar, and a typewriter to enter the events into each day block, one at a time. It took a month to make the twelve pages.

We looked at many options for calendaring automation packages. We realized early on that we were looking at a facilities scheduling problem, not a calendaring problem, because the consensus of opinion was that other (non-room scheduling) events were not going to be put on this calendar.

After auditioning several programs for the period of one month each, it was determined that automating this process was not feasible at our parish. There were some hefty requirements. The program had to allow room-scheduling conflicts to occur, so that the staff could continue to go through the process of compromising about room use.

Additionally, the program had to generate a block-type calendar, which would allow all events for a month to be placed on one piece of paper. This was nearly impossible. The best we could get in an off-the-shelf package was the ability to make a "compromised" block calendar. If there were a great deal of events on a particular day, the calendar would "spill" onto a second, or even a third page. This made the paper-printout calendars very difficult to use, and was the principle reason why the program was discarded as an option. Our people do not work in offices, primarily, and need to go to meetings outside of our system, and without the benefit of computers, so this paper-calendar was of key importance to them.

While it may seem that this is a failure, it really is not. It was determined that by automating the process of making the actual calendar using desktop publishing software, and maintaining the manual process doing the scheduling, a happy medium was achieved, which satisfied the users of the system.

This shows that with cooperation, a satisfactory resolution can be achieved.

7* *If this is a paper-based process, are there any special forms used?*

The presence of special forms is very important, because it shows some of the unspoken or unwritten nature of a process.

8* *What other processes does this process affect? How does this process affected by other processes? Can you forsee any new processes in the future that this process could affect, or be affected by?*

The answers to these three questions are important, because they help to give the overview of what will be affected if and when you implement a system. There is a codified system, called a "Work Flow" document, which can help to illustrate for both the designers and implementers of a system how a process works. Misunderstanding of how a process interacts with other processes is a classic example of a stumbling block.

9* *How long are you preparing to live with the automated system which you put in place.*

Is this an project seen as an interim solution, is it the permanent resolution to a problem? The answers to this should usually be resonant with how much money and time is being put behind implementing it in your organization.

10* *What are your real expectations of the features/cost of this system?*

Ask everyone this, because you will find that adding items to the wish list is important at this stage. Be certain to make no promises about features or improvements, but do listen to what people have to say.

11* *What is the time frame for restructuring this process as an automated process?*

Once again, this should be resonant with the answers to questions 9 and 12. A bigger project should be ventured into prudently.

12* *How much money is budgeted for the initial capital expenditure?*
& *13*

How much money is budgeted for this automation project this year, and annually for the next five years?

These are tricky questions, as many people do not feel comfortable asking about budgeting or financial information. It would not be appropriate to ask everyone this question, but it is important that you have some financial information. These two questions are especially difficult because the background is rarely in place to project budget information. Numbers are not key at this point, but try to have an honest appraisal of the scope of your project. Beware answers involving excessive vagaries or avoidance, and especially of the "not too much money" type. At this point in evaluation, these are dangerous answers from people who may not fully appreciate what they are getting involved with. This will likely effect the amount of buy-in to do the project "right", versus doing the project "on the cheap". Because the level of funding in non-profits varies significantly, there is no "right answer" to these questions. All charitable (as well as for-profit) organizations need to do everything with an eye on economy, but beware of excess, as it may undermine the success of the project. There is a world of difference between intelligent economy and suffocation.

The latter of these two questions is especially important, because it opens the mindset for the idea of future support, a very important factor in any project. Planning support costs will make a huge difference in the satisfaction in the project down the road, it may also avoid excessive cost-cutting in the implementation phase.

Summarizing the Information

Once you've asked these questions, It is a good idea to have something a basic framework to work with. Collate and summarize each of the answers to each of the above questions. While there may not be parity, your questions have given you a "top level" plan of the process. If a particular individual's answers confuse you, or if you feel that there are holes in your investigation, it is crucial to go back and clear them up at this point, because this is the genesis by which the rest of your project will stem.

Keep in mind the previously described concept of the "work flow document". For some types of systems, this level of design may seem to be overkill. It depends on the project. Some people have a tendency to over-document, for others it may be necessary to design a plan for information flow that is very basic. The key to all of this is that you, as the system developer, have a firm harness on the process, and that you can communicate this message effectively to management and to the vendors, when you get to that step.

RFQS, SPECIFICATIONS, AND DEALING WITH VENDORS

In time, it will become necessary to begin looking at making purchases for network equipment or services. Dealing with computer equipment vendors is a lot like buying an automobile. Sometimes it can be like buying a car from someone who knows less about cars than you do. Computer dealerships run the gamut from specialty "system integrators", to huge consumer electronics or computer chains; the "big-box" store. Meeting with as many vendors as possible is integral to this process. Be aware, however, the beginning of this search will mean the end of your ability to work uninterrupted by the telephone for the next six months!

While personalities vary, vendors can be like sharks (Sorry, guys). They are basically good guys, and can be the good friends of a system manager, but first and foremost their job is to sell things. They have rents to pay, and families to feed. While your performance is based on consistency of operation, theirs is based on consistent levels of sale. Trust me, vendors deal with people like us everyday, and they can smell blood.

Your best strategy is to use the vendors to your advantage. One thing about being a vendor, especially a small vendor in a large market is 'banging the drum' about what can be provided. People who sell systems sometimes have a package, a set of services, which they try to offer to many people, in some kind of modified form. This works for them, because support channels are standardized and predictable, a level of physical inventory and knowledge can be maintained, and the economy of scale is better for the vendor. Packages can work for you, too, as the support turn-around time is usually faster.

In almost every circumstance, the big-box store is not a good choice for businesses or commercial interests. The post-sale support is usually very poor, and the sales staff knowledge of the product line is limited to the consumer-oriented products that the particular store sells, *at that particular time*. The memory span of the sales staff is terribly limited, and they never have any fore-knowledge of what may be available in the future. Extensive information about a product is not usually available, as is information about other options in the product line beyond what that particular store sells. Because these chains operate on quantity rather than quality, the latter in these stores is frequently sacrificed for the sake of economy.

Typically, you want to go one of the other routes when looking at vendors for your organization. These options are either the local computer store, or the mail order catalog. I have relationships with both types of retail outlet. You may also find that dealing with vendors is not consistent. Certain types of purchases are better made with certain types of vendors. There is no "one" avenue that will completely satisfy all of your varied needs.

For example, When I began building our network, it became apparent that we were going to need to buy in the neighborhood of 50 computers for use by the clients. This was a purchase, at the time, of around $60,000. There were certain things we needed to have, like a credit account sufficient to make the purchase, warrantees on the equipment, and the ability to get these machines configured with operating systems and hardware to our specification. I interviewed six vendors.

First, I interviewed the consulting company from whom we purchased our servers and networking equipment. I then interviewed two local PC shops with which I had done previous business, although not with this organization. I also interviewed reps from the catalog vendor with whom the organization had a good relationship, a Tier-1 computer manufacturer, and finally a consumer electronics retail chain (a big box store).

What were the results? I had an excellent experience with the local consulting company. They dealt in one brand of computers, and told me that they could put together the right package for my needs. The consultant gave me a binder, indicating their scope of work for this project, and price points. Although their prices were about 1/3 higher than my budgeted cost for these machines, these machines were of recognized brand-name, and the firm was to provide some additional professional services, such as setting up the computers with virus protection and the network. We had previously dealt with this firm, and sufficient credit was established, so logistically speaking, the purchase would be the matter of a faxed purchase order. They indicated that they would be able to get my equipment delivered and installed by our deadline.

The Local PC vendors each gave me a computer to try out. I was able to get, within a day or two from each, exact pricing and specifications lists. I needed the documentation to be in a particular format and layout, which one vendor was able to provide for me up front. Both were willing to extend 30 days of credit to us, with the appropriate documentation. They indicated that should I place the order within two weeks, they would be able meet my delivery deadline.

The representative from the catalog store asked me about the desired price point, and sent me various sales flyers for computers that fit. The representative was unable to format my specifications in the binder, but she sent me everything I needed in terms of technical information so that I could put it together myself. The information that was not readily available, the representative sent me by fax the next day. On her suggestion, we began the process of setting up credit. They indicated that for an order of this size, that the equipment would need to be drop-shipped by the manufacturer, and would thus be delivered on a palette by a trucking company.

The tier-1 manufacturer routed me repeatedly to a voice mail box. Because we fell into the classification of "Education and Government" type buyers (meaning that we are tax exempt), we were told by the usual representative to deal with the company's regional representative. Although the regional representative didn't return my phone calls in time to become a real part of the evaluation process, I was able to get pricing and availability information from their website, and as such, I compiled a partial binder for this option. As the commercial grade equipment offered was well out of the price range, no further action was taken.

Finally, I went to the consumer electronics chain store. I went on a Tuesday morning around 10 O'clock because I felt that this would be better than the afternoon, when the crowds usually appear. There was no one to be found in the computer hardware department. After walking around for 20 minutes, I found someone in the television department, who was able to find someone to help me. I was shown an array of consumer-oriented computers, many of which were well within my price range. When I asked for technical specifications other than those printed on the sales material, I was directed to the manufacturer's website for more information.

Oddly, as I am not a fan of chain electronics stores, I was intrigued by this last option. I was hopeful, because we had a choice of name brand computers, some well under the projected price point in our budget. I projected that we might well be able to put 10 to 15% more machines into the installation.

The problem began with the software. There was no "operating system" installation disc included with any of the computers that we were looking at. They all included "system restore" discs. Some of these computers were running lots of extraneous software, and it appeared as if we would have to reinstall, then uninstall all of this stuff every time a computer needed to be overhauled. This was not an insuperable problem, as we had the ability to do network installations, or use an automated installation from a master CD.

The real problem was making the actual purchase. When I told the salesman that we would need 50 or so of these machines, I was told that it would be necessary for us to fill out a "commercial credit" form, and to deal with the ordering department which deals with business clients. Business clients had a counter on the other side of the store, so I went over there and inquired about credit. I was handed a form, and told that this particular retailer did not issue credit. The "home office" needed to be contacted. The clerk allowed me to call from the Commercial Relations desk. I spoke with a representative, and was told that a commitment of $2000 per month for the period of 1 year was required to establish credit. I took the information, thanked the salesman who had helped me, and left the store.

Ultimately, we decided to go with one of the local PC shops. Why did we do this? We made the decision based on two factors: Quality of the product, and the fact that the firm was based locally. While conventionally speaking, this may not appear the most prudent choice when compared to the array of other venues that we considered, it has worked out exceedingly well. These computers basically have worked flawlessly over the last 3 1/2 years. There have been minor problems, but one definite advantage exists: It is exceptionally easy to fix these computers, as they are basically generic. The power supplies, motherboards, RAM, keyboards, and other component parts are interchangeable, and are readily available. We can stock standard cases, hard drives, CD-ROM drives, and thus, eliminate the key problem of Tier-1 equipment: Custom Parts. The majority of the component equipment in a brand-name PC is of proprietary design. While a lot of this has to do with engineering a competitive product, the custom nature encourages you to work with their service providers.

Here is a chart indicating various advantages and disadvantages of different types of sales providers. In your search for a vendor, your relationship with the people is as important as the product that you are purchasing. Computer vendors come and go, so it is a good idea to get references before choosing of these options.

VENDOR TYPE	ADVANTAGE/DISADVANTAGE	POSSIBLE ACTION	WHAT TO BUY
CATALOGS, MAIL ORDER AND ONLINE ORDER VENUES	Inherently less personal, because impossibility of face-to-face communication. Cataloges are good because they can offer a great deal of variety in product avaiability. Actually receiving the merchandise, and Returns can be more difficult, as shipping is an added concern	You may prefer to deal with real face-to-face people. The catalog route is good if you can get a specific, dependable representative.	General Supplies for Printer and System maintenance Operating System and Application Software Backup Tapes
PROFESSIONAL SYSTEM CONSULTANTS	A good option for complex projects or those projects which require integration. Consultants may have an objective viewpoint and a clearer, different approach to solving a problem	Consultants are good for projects with specific goals and scopes. You need to do a great deal of planning before engaging a consultant. They can get expensive, if left to their own devices	Servers and LAN Equipment Network Designs Whole-Solutions packages, such as Firewalls, Antivirus Systems, and other protection schemes
LOCAL PC VENDOR	Usually very friendly, locally based, knowledgeable company can be a real ally when maintaining a network. This is your "meat and potatoes" venue.	A good option for client PCs, because these shops are set up to build client equipment at a reasonable cost and of acceptable quality for this purpose. Servers can be dicy.	Client PCs Miscellaneous parts
BIG BOX STORE	Lots of demo equipment.. Lots of brand choices, but in a very narrow range of various manufacturers' product lines. Sales staff will not have a lot of expertise in any area. May be a comfortable environment, as you have no doubt dealt with one for personal purchases.	Not a good venue for any component of a networksed system. Computers tend to come preloaded with a ot of software, most of which is useless to your purposes. Service, in general, is of inconsistant quality	Digital Cameras A/V Equipment May be a good place to demo some different options

FIGURE 1.5: VENDOR ATTRIBUTES

Never underestimate the value of your vendors. There is usually a stigma surrounding dealing with people who are trying to sell you services or products. Vendors are very important, especially when it comes to larger scale projects. They cannot make your decisions for you, but they can provide varied options, if you know how to ask. It is very likely that in your situation, that your local vendors have been down these roads before. Get lots of detail, as much as possible. Salesmen typically love to talk about their products, and are usually perfectly willing to draw out surprisingly detailed designs for your system. They usually have lots of spec sheets and brochures, which are laid out with managers in mind, not technical people. Check with lots of vendors of different sizes and types, and get their opinions on how to set up your systems. This information will help you to define your specifications, especially for topics which you may not have lots of expertise.

Bear in mind, though, that these people should not be the only basis of your knowledge. Sales staff and marketing people put a spin on products, instinctually, as part of their jobs. They typically do not deal with post-sale issues, and problems. The glossy brochure will list a lot of features of a particular program or system, but may not indicate how well it works with other systems, particularly in cases where those systems are not made by the same manufacturer.

THE NEEDS ANALYSIS PROCESS

Needs documents are a good tool to use when beginning to plan a system. They should be appropriate to the size and number of people and groups involved with a project.

A Needs Document is supposed to vocalize base requirements and desired features of a proposed program. It is supposed to be able to be used by the decision-maker to determine which of the optional choices is best. It is based on a numerical system of prioritization of features. Needs documents are important, because they help to guide the search process, and guide the evaluation process. Sometimes, this process is quite complicated, other times it is not.

An example of a fairly simple decision on our network was the decision of topology. These decisions were made based on accepted standards, which have been adopted almost universally, in almost all industrial and commercial applications of the time when this system was installed. There is almost 100% compatibility, both backward, and for equipment made in the foreseeable future. For our purposes there is virtually no downside to this type of a system.

The Calendaring System

Converse to the relatively easy decisions regarding the network, which took two meetings to hammer out, the question concerning our master calendar was far more intricate.

Most parishes, especially larger ones with multiple meeting rooms and groups, eventually realize a need for a master calendar. Presently, although software exists to handle this function, the calendar in a parish is typically of a large block type, on paper, located in some central area. The calendar serves two principle functions. The first is as a tool for communicating what is happening at the facility. The second purpose is not so obvious.

A master calendar can act as an objective arbitrator in situations where scheduling conflicts have occurred. Rules and protocols can be established. In situations where otherwise fighting or ill will amongst groups or individuals can arise, a master calendar can help to sidestep or resolve.

Our decision regarding the Calendaring software package was somewhat more in-depth. First, although it was a new automation project, it was not a new system. There was a lot of buy-in to do this process automatically, but there was not a lot of experience on our staff with the ramifications of an all electronic system.

The first thing that I did was to have a meeting. Our professional staff (those who use the calendar) assembled. I brought in an impartial third party (a member of my technology committee who is not part of the parish staff). He conducted a conversation amongst all of us. I participated, but only to answer technical questions that arose in the course of the discussion. The discussion lasted for about two hours, and covered lots of things having to do with the way that our calendaring happened. It was amazing, because I do not feel as if there had ever been a real discussion about our calendaring process before this.

I then took the notes about what people had said, made a survey and delivered it to every staff member. I filled out a survey as well, because I too would use the calendar program. Please note also that there are no issues on this survey other than ones which were specifically raised through the course of the discussion.

PART ONE—PLEASE READ ALL OF THE DIRECTIONS

All of the following statements were made at the last staff meeting concerning scheduling. Please rate each of these statements comparatively to each other on the list. Assign 1 to the most important of the items on the list, 6 to the least important on the list, in your opinion. PLEASE READ THROUGH EVERY ITEM THOROUGHLY PRIOR TO RATING ANY OF THE ITEMS ON THE LIST. IF YOU NEED FURTHER EXPLANATION ON RATING THESE ITEMS, PLEASE ASK ME.

The ability to view all room reservations for a given period of time at once   _____
on the display screen

The ability to print out a single room reservation schedule for a given period   _____
of time

The ability to print out all reservations over a period of time   _____

The ability to schedule multiple groups in one room   _____
(the ability for the system to allow scheduling conflicts)

The ability to have automatic re-occurring events involving room scheduling   _____
(eg. "I want the system to automatically schedule this meeting every second
Tuesday of the month for the next calendar year")

The ability to interact your personal schedule with a Palm-operating system   _____
enabled device

PART 2

Please read through the following statements (which are the same as the above statements). Please rate each on a scale from 1 to 10, where one is critical or very desirable, 10 being useless or not desirable.
For this section, please rate each one as its own entity, and not in relationship to the others on the list.

The ability to view all room reservations for a given period of time at once   _____
on the display screen

The ability to print out a single room reservation schedule for a given period   _____
of time

The ability to print out all reservations over a period of time   _____

The ability to schedule multiple groups in one room   _____
(the ability for the system to allow scheduling conflicts)

The ability to have automatic re-occurring events involving room scheduling   _____
(eg. "I want the system to automatically schedule this meeting every second
Tuesday of the month for the next calendar year")

The ability to interact your personal schedule with a Palm-operating system   _____
enabled device

After having received all of the surveys back, I compiled the numbers, and made a prioritized list of the above six issues. Following is the list of their results, as tabulated. There are 13 people who were polled, and their results were listed in columns, starting with the first person, and ending with the last. Each person's results were weighted equally, to figure the averages.

SURVEY RESULTS

SECTION I.

STATEMENT	1	2	3	4	5	6	7	8	9	10	11	AVG
No. 1	2	1	1	3	2	1	1	3	1	2	5	2.000
No. 2	3	5	4	5	4	4	5	4	4	5	2	4.091
No. 3	4	4	2	4	5	2	2	2	2	4	2	3.000
No. 4	5	2	5	2	6	6	3	5	5	1	3	3.909
No. 5	1	3	3	1	1	3	4	1	3	3	4	2.455
No. 6	6	6	6	6	3	5	6	6	6	6	6	5.636

SECTION II.

	1	2	3	4	5	6	7	8	9	10	11	AVG
No. 1	2	10	10	9	8	10	10	8	9	4	5	7.727
No. 2	2	6	6	9	7	3	6	7	8	5	10	6.273
No. 3	4	6	9	9	7	10	10	8	10	6	10	8.091
No. 4	6	10	5	10	5	3	8	4	6	10	8	6.818
No. 5	2	10	8	10	10	6	6	8	7	10	8	7.727
No. 6	1	1	1	6	9	6	1	2	1	1	6	3.182

FIGURE 1.6: RESULTS OF NEEDS ANALYSIS

There were a number of baseline requirements that were added to this list of variables. A provision for an ongoing support agreement, which could be renewed at our option, for at least five years after the purchase date. In addition, individual user accounts, web integration, and the ability to work with our operating systems, backup systems, and network were required. I added these to a big list, with those baseline requirements at the top. I then prioritized the users' requirements, and added them to the list.

After all of the information has been compiled, a document summarizing the findings should be produced. This document is a vocalization of the varied requirements and desires that this new system should be able to accommodate.

CALENDARING SOFTWARE

<u>I. SCOPE</u>

This proposed system is to be defined so as to provide an adequate replacement service for what is locally known as the "master calendar". Upon successful implementation of this project, the existing calendar function is to be superceded with this electronic system. This system will serve all network users, and will be available from all network-connected systems.

This system will serve the person who defines the list for the weekly bulletin (to determine what is occurring during a given week, by combining the resource calendars for each room, and using this information to build a weekly "event" list). This system is to be used by the staff to book shared resources like rooms, audiovisual equipment, and events, in addition to their own schedules.

This system will manage the entire event and room calendar.

REQUIREMENTS

This system should incorporate (1) email, (2) individual scheduling, and (3) group [resource] scheduling functions. This system should also allow for the ability to manage documents relating to resource usage. The system should incorporate on the server and the client side the email function and calendar management functions into one application. The system should allow for the addition or deletion of resources, whether these resources are rooms, AV equipment, or other equipment or resources.

This system should allow different tiers of users. There should be the ability to view resource schedules for all users. There should be the ability to view individuals' schedules if a user has access. There should be different tiers of users who can add/change/delete their own resource reservations, and tiers of users who can add/change/delete all room reservations (resource managers). In addition, there should be a definition of user class who can add/change/delete all of the resource calendar items, for maintenance and conflict resolution purposes.

The system should allow password protection (either its own, or through existing user management) for individual users, and a guest classification for users who do not have accounts on our network.

CLIENT SIDE SOFTWARE REQUIREMENTS

Client software must be able to run acceptably in terms of speed and usability on Windows 98, ME, and NT 4.0 systems. System must be able to be simultaneously available to any network user who wishes to use it.

The system must be of recognized brand and manufacturer. Adequate technical Support and user/administrator training materials must be available for this product.

SERVER SIDE SOFTWARE REQUIREMENTS

The server-side software should be able to run either on its own Windows NT server, or on its own server which can be integrated into the existing Windows NT network architecture, using TCP-IP to connect to the system. The server software should be able to integrate with Internet email, local email, and group distribution lists for email distribution. This software should handle some form of document management system to allow reference documents to be accessed by users of the system.

The system should be able to handle 50 to 75 users acceptably in terms of response time.

TIME FRAME

(Time Frame Information)

II. THE PROCESS

DEFINITION

- determining needs of users

- Defining needs document

- Reviewing software options

- Determining which package to use

DEVELOPMENT

- Obtaining the software

- Setting up server-side software, and the server.

- Setting up user/email accounts on new server, and connecting the server to the internet email transfer.

- Setting up client side email/calendar software and testing configuration

IMPLEMENTATION

- Training and introduction of new email software to all users

- Retiring old email server

- Training and introduction of new Calendar software to Professional Staff

- Training and introduction of new Calendar software to Administrative/Secretarial staff

MAINTENANCE

- maintaining software and server, implementing patches and upgrades

- periodic reviews to determine if software is meeting needs

III. ADDENDUM

TERMS AND CONDITIONS
(Terms and conditions information)

TIME FRAME
(Time Frame Information)

CONTACT PERSONS
(A list of Contact people)

DEALING WITH VENDORS

The RFQ

September will be here soon, with cooler weather, deep fall oranges and yellows, and…students. The repair of our school's PA system is nearing the end of the second week. There seems to be no end to the line of problems with this system. Much of the wiring is damaged by the new construction. The deteriorating console is from another age, when people still feared the Red Threat, and "Transistorized" was something to advertise. Who knows, some of the speakers may even have squawked out "Duck and Cover!" The two guys from our sound contractor's Service department talk of interference, "cross talk", and impedance problems. I am told that everywhere in the old building the announcer's voice is indistinguishable, and that many of the speakers are no longer useable. I see two of the installers outside pulling gray wire along the outdoor courtyard, using the 9x9 in-ground tiles as a crude ruler. I wonder when they will finish, and I wonder how much it will cost.

August 25, 2000

A Request for Quote is a document you can send to a system installer or vendor. It asks for a quote to be returned for a designed system, and what needs to be included in said quote. Be aware, you are also sending a signal that you have a serious intent to at least consider using this vendor for your project. Read: Don't send RFQs to lots of extraneous companies; there are better ways to get information.

RFQs are crucial when doing big projects. They let your management groups know that you are serious about doing this project in a professional and thorough manner, and that you have consulted others about the real needs and goals of a project. They also let the vendors know this, which puts your relationship with the potential sales representative on an equal level. A well written, tight RFQ is a signal to the vendor that you do know enough about what you need to not be talked into a host of extraneous options. This is not to say that the vendor will do this, but it does help to clarify your position.

The other side of the RFQ is that it forces the vendor to produce something tangible to return to you. Typically, the vendor will call on you to look at the existing system, or other aspects of the job environment. This is especially true in jobs involving physical modifications, wiring, or hardware to be installed at your premise. Even in the case of a software project, you will find that they will want to see what your system has in place, and to try to gauge your level of expectation.

The RFQ for our calendar project was one page long, and was divided into three sections.

The first section listed our basic goals for an automated calendar system, and summarized the process by which the calendaring was done at our parish.

The second section was the above list of desired features, beginning in priority from the top. The first baseline requirements were asterisked, noting that they were requirements, and then features listed by priority.

The third section listed information about the time frame for the implementation of this project, and a projected time line for key aspects, such as meetings, training, etc.

Other RFQs may specify other types of information. A minimal RFQ may include the following:

- A cover page with the date, the contact information and organization name, and the project being described within

- An introductory paragraph or two describing the background of your organization

- An introductory paragraph describing the system which is to be quoted

- A set of bullet points describing the desired results of the overall system to be quoted

- Numbered subheadings, containing paragraphs with descriptions for each of the different discrete aspects of the system, and bullet points outlining their desired results

- A section, which outlines terms and conditions of the proposal. This section might specify any standards adherence or the level or format of the documentation which will be required in the proposal.

- A paragraph outlining your selection criteria. This paragraph might include the name of the individual or group which recommends, and the name of the person who makes the decision. It may also include if known, the projected date or time frame for a decision. Finally, it outlines the method, time frame, and format in which a response will be sent to the vendors who do as well as do not receive the project.

You sent out the RFQ, now What?

Well, now you wait. More than likely, the RFQ will signal to the vendors that you are serious about going forward. There should be at least three to five bids on a project your organization considers to be significant, in scope or cost.

In about a week, you will begin to get calls from sales representatives, and depending on the type of contractor, perhaps even the head of the company. A responsive vendor or contractor will want to meet with you personally before submitting a bid. Bids are a complex and potentially dangerous arrangement for any contractor. For you, the bid may be figuring out who is giving the most for the least amount of money. For the vendor, it is figuring out the least expensive way to get the job done, while remaining in line with your specifications. The vendor's ability to estimate accurately can make the difference between getting the job at all, and going into the hole on the job. Some vendors may even try to get you to talk about the other bids. This can work both for and against you. It has been my experience that giving them a general idea of the budget is not a bad plan. Some sales oriented people say you should never give away the amount you have to spend, but too much secrecy forces the bidder to take shots in the dark when doing the bid. What you end up with is wildly different bids, which makes objective comparison difficult

This being said, you should never give away the position of the other bidders. There is a tendency to become friendly with the sales people. You may build a relationship with these guys as they come out to look at aspects of the job. They may ask about the other bidders. It is unfair and unethical to give away information about the other vendors, as it undermines the level playing field that you have worked to build with your RFQ. The most information that you would want to give is how many other vendors have bid on the job, and possibly, how many of the other bids have come back. Comparative statements about bids, and even disclosure about the other companies involved can potentially cause collusion, or other dishonest practices in the bidding process.

Eventually, one of two things will happen. Either a bid in some form will come to you, or the bidder may back out of the bidding process altogether.

The Bid

Bids vary depending on the experience and culture of the bidding company. I have received half page bids. For the same projects, I also received binders of information along with contractual agreements, terms, and component product literature included. The bid may look very formal, or very casual, depending on the contractor.

Judge the style, quality and comprehensiveness of the bid. In all likelihood, the job will be completed in a similar style. If the bid seems expensive and stuffy, the job or post-sale support will probably end up being the same way. If the bid was carelessly thrown together, or sloppily faxed to you, there is a good possibility that the job too will be done unprofessionally.

Avoid altogether bids that clearly do not follow your guidelines of bidding. Guidelines should be clearly stated at the end of the bid. If there are glaring differences between what you asked for and what you got back, this will probably be resonant in the actual job done.

Look at all of the data given to you. A bid should include all aspects of what you asked for. Look for these sections, specifically, to be present:

- References from other local firms or organizations, specifically from installations using similar equipment or designs that you are asking for. Do not hesitate to follow-up with references, and ask if it is possible to see their some of their other work.

- Post-installation or Post-sale support agreements, with provisions for renewal for an appropriate amount of time beyond the warrantees of the actual Components of the system. This will help to ensure that this vendor or Contractor is around for more than a year. Consistency, in all types of Relationships, is usually a good idea.

- Training options regarding the installed equipment, specifying in writing the level of training included in the installation price, the optional training available and pricing for this, for whom the training is available, and when the training will occur. When our HVAC equipment was installed, part of the agreement was to go through the entire system with an engineer and a camera operator. The camera operator taped the entire walkthrough, for later reference.

- Documentation and certification which will be provided, as appropriate.

- Any obvious aspect of the system which is outside of the realm of the contract. For instance, we deal with a sound engineering company that has done several large-scale sound system installations for us.

Up front, they put in the bid that "Electrical work is outside of the scope of the bid". This is printed in bold print on the front of the bid, In the space where all of the component pricing and information is described.

- A listing and pricing of parts used to complete the job, along with brands and model numbers. This may be somewhat tricky to get, but it is not abnormal to ask for in a bidding situation. Some vendors are hesitant to give away their "special blend", which is understandable. Remember that bidding costs a company time. It is more involved than simply checking a price quote. Respect this, but ask for anything that you can get from a vendor. Even if you do not use that particular vendor, it will give you a good idea about their direction to a solution for your problem.

- A list of telephone numbers, and local addresses for the vendor should be present somewhere in the bid.

- Find out the terms of the contract, and if a contingency is required. Is the contract a fixed price, or can prices vary depending on certain circumstances beyond the control of the vendor? Contingencies are typically the realm of building projects, which can increase by hundreds of thousands of dollars due to unforeseen circumstances. Very few network projects require any contingency money to be put aside, but depending on how the contract is written with regard to variances in the price from the quote, you may want to have a backup plan in your own budget process for problems with the installation.

Terms

Once you have signed a contract, the next logical step is the credit terms. There are very few circumstances where you want to pay for a project entirely up front. Just like having a house built or buying a car, having some amount of money outstanding puts you in a position of greater leverage with the contractor. Many computer-related projects are spread out over weeks or months. Networking contractors are no different than building contractors. Once the check is cashed, some may be inclined to move you to the background and do other more pressing jobs.

Unless there is some extenuating circumstance, You probably want to pay no more than 1/3 of the project up front, leaving at least 1/3 for a period of time after both parties (you and the contractor) agree on the completion of the project. Incidentally, this arrangement will also help them. In the event of an increase in the cost of the project that is not covered under the contract, it is a lot easier to bill the variance (or issue you a credit if the opposite conditions exist).

THE DEMONSTRATION

Once you have chosen an acceptable path, it may be a good idea to introduce the proposal to key people in your organization. Part of this is to bring others into the loop on your project, making them feel as though they have a hand in the decision-making process. Another benefit of this is for review. Demonstrations, while sometimes like test-driving a car, can be a good check and balance system. Any fundamental flaws will appear in a demonstration of a product, and a shoddily built product will usually not demo particularly well. Having said this, a demo can be "110%" of a product. A skilled salesman knows his product, and can make it shine. While a demo will not make a "silk purse out of a sow's ear", it may minimize or hid weaknesses in a product.

Demonstrations should be like automotive test-drives. Orchestrate a demo using a one hour session, and allow time for questions. Demonstrations should always be in as quiet, comfortable location as possible *without external distractions*. Never hold demonstrations in offices, or at someone's desk, as this will undermine your ability to concentrate on the product demonstration. Many product manufacturers and vendors want to do the operation of the demonstration themselves, for an audience. Be wary of any product demo that seems to be scripted. Make the demonstrator allow you to try the program out, and always try to do things which the product literature does not advertise.

Because you are testing for usability, be aware of performance and handling. If this is a software product, make notes of the unquantifiable aspects of the program. Spec sheets and product literature will enumerate the features of a product, but the demonstration will illustrate the execution of the features. Responsiveness in most any piece of production equipment or software program should be smooth and quick. If you are aware of the product itself, it will soon become a weapon in your organization, rather than a tool.

PILOTING AND EVALUATION

Prior to a real purchase or rollout of a product, it is a good idea to evaluate it in your organization whenever possible. Product vendors, especially software vendors, should be amenable to this. An evaluation allows your network support staff (or yourself) to actually set up a microcosm of this program. Stumbling blocks in the installation process may become apparent.

Evaluating certain types of products may be more difficult than others. Installing any product, hardware or software, which affects the overall system should be avoided. If possible, a small "staging area" should be set up, initially, to make certain that the product will work with the existing system.

Networks are a unique arrangement in this way. Like many products, even a small network is made up of a host of thousands of different manufacturer's components. Despite the existence of external standards, the marriage of many vendors and manufacturers to make a "system" is never easy. Even if your equipment is all manufactured under the same brand, every computer contains thousands of different components. Even newer generations of products branded the same will contain different manufacturers' products.

In this way, software is much more precarious than hardware. With literally thousands of people behind software products, a symphony (and sometimes a cacophony) of products may or may not work together. This is the real explanation behind why computers crash so often. Any computer system is like a house of cards, with modification going on constantly. While it is unlikely that one product will bring down an entire system of computers, a product integration scheme which is not adequately tested or designed will make future trouble-shooting network problems very difficult.

Compared with an automobile or an aircraft, a networked system is a loosely connected arrangement. No one vendor has the capacity to test and configure every possible configuration in the real world. A network administrator has the responsibility to make sure that all of the products work together, and that no individual product unsteadies the rest of the balance.

SUMMER: FRUITION

○ ○

"...He who sows sparingly will also reap sparingly, and he who sows bountifully will also reap bountifully"

—(II Cor. 9:6).

Imagine an empty room with no windows, no ceiling, and a concrete floor, roughly the size of a small bedroom. This room is buried in the basement of a brand new building. Even to many who planned this building, this place is forgotten or unknown. It is not the showpiece of this building. It is hidden in the back corner of an oddly shaped storage room in the basement. There is no sign on the door. There is no indication of its existence.

As I stand here, I feel the potential screaming amongst the four walls of this place. One day soon, it will be a very busy place. It will be a system room, the base of operations of a hundred-thousand dollar plus system. I am 23 years old, and I stand here in blue jeans and a T-shirt with three people, the youngest of whom is twice my age. I have just completed the best interview of my life to this point. I have never managed a project like this before, but these people are looking to me to make decisions that will affect this place for many years to come. I have been hired to be the expert. I have been hired to deliver this system to these people, and I have been hired to be responsible for what will happen here.

August 5, 1999

DEVELOPMENT

Up to this point, a lot about the purchasing of the system has been discussed. Before any purchasing decisions are made, however, technical decisions must be made And plans must be drawn. Computer systems really are about doing three things quickly and reliably. These three things are processing of data, storage of data and communications. Individuals run programs on their computers to access data on the server or use programs running on the server, in a client/server, and host setup respectively. The server handles some or all of the processing of the information, showing the results to the user on the screen of their computer. The Server handles all of the data integrity, security, and backup functions.

Most network operating systems require you to apply to each client and server system a unique name. Some systems do not require names for client computers, but do require them for servers. I think that the best naming convention that I have seen was in college. The college had a computer-science lab containing 14 workstations. Half were about ten years old at the time, the other half were much newer. The seven older workstations were named after the seven deadly sins, the newer machines were named after the seven saintly virtues. I wonder if there is any bad karma in using a computer named "Sloth", or "Envy"?

How many servers you have depends on the extent to which your users will be reliant on the system to do their work. Usually, you want to have at least two servers, for the sake of redundancy. Many network operating systems allow redundancy through the use of multiple servers. This allows things like user accounts and rights, and other pertinent information about the network like the list of computer names, to be backed up in real time on both machines.

While some servers look on the surface just like client workstations, servers are really very different than client machines. Servers are about doing one or two things really well, and really reliably. Servers are about redundancy, and spending as much time as possible doing these one or two things, as opposed to being 'down'. You want to try to isolate servers physically and logically depending on their intent and purpose. While a client machine is optimized for user-oriented performance, a Server is optimized to communicate and process well. Client computers are usually designed to run the greatest range of programs, and to do the greatest range of applications. Reliability is usually important, but not crucial, as the user can easily restart a crashed machine.

For a server, reliability is really the key point. Your network is only as strong as the weakest link. As good as the wiring or as fast as the client workstations may be, the network will not be functional if the servers lack stability. Unfortunately, a reality of life is that the working network is invisible, while its failure is criticized loudly and often, and worse, remembered for a long time.

Preempting weakness that can cause failure is your responsibility as the manager of a network. In most all situations, It is a good idea to shoot for higher quality rather than flashy perks when making decisions about a server. Extraneous hardware, such as sound cards, modems (unless necessary for operation), and fancy video cards should be avoided. Brand name hard disk drives should be used, in redundant pairs with some kind of data synchronization scheme, such as a RAID array. RAID (Redundant Array of Independent Disks) is a scheme where several physical drives are linked together, and thus form one big drive. In certain types of RAID setups, if one drive fails in a RAID set of drives, no data is lost. This should not be regarded as backup in a traditional sense, but it can make a difference if and when a drive fails. RAID arrays used to be expensive luxuries, limited to the most advanced server setups. The advent of IDE RAID cards has ended this, and these drives and cards can be installed for a reasonable price.

In addition, brand-name components, like RAM and network cards should be used in servers. Generic network cards, especially, are prone to problems, and should be avoided. The cost of the brand-name components is not significantly higher than that of a generic counterpart, but the cost to support the latter is significantly higher.

Beyond quality and redundancy, you will need to analyze your needs and resources. A list of introductory services you may wish to provide should be made.

> *Data Storage (Private for each user)*
>
> *Data Storage (shared for all network users)*
>
> *Program Installer and Utility Storage (Publicly accessible)*
>
> *Internet Email for internal users*
>
> *Base Services*

INFORMATION STORAGE

These three items require hard disk space, and backup services. They also require that users have user accounts and authentication (Meaning that they have passwords). Whenever possible, you want your users to have their own passwords and user accounts. Users should each have a unique account attached to their name.

How much storage space is necessary will vary with the application and environment of your network. For general information (data) storage, between 100Mb and 500Mb per user should be sufficient.

Server Setup

First, before you look at user storage, lets look at how the system should be set up. There should be two Ethernet cards, of different manufacturer. One should be set up with network settings for the network, the other should be set up with a dummy setting, so that it can be quickly reconfigured as the primary, should there be a problem at an inopportune time.

There should be a standard video card, and modem. These two items should work properly, and should be tested periodically.

There should be a working CD Writer and Floppy drive, for data transfer purposes, in case of some problem which precludes network access.

There should be two of the exact same drives for the system, which are set up as a mirror. A mirrored drive makes an exact copy on the backup drive, so that if there is a problem with the first drive, there is a backup on the second drive.

There should be two of the exact same drives for the data drive. These should also be set up as a mirror. Additionally, you should keep on hand:

- An extra Power supply

- Extra RAM memory

- An extra System-type drive

- An extra Data-type drive

- An extra Ethernet card

- An extra video card

- An extra lithium battery for clock backup

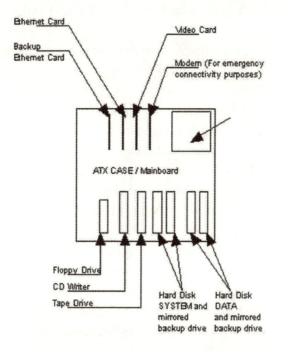

FIGURE 2.1: TOP-VIEW OF A SERVER

Well, back to the data calculation. How do you calculate how much space you need? Well, lets say you are going to a lot 100 Megabytes of space to each user.

Number of Users x 100 Mb = Total Drive Space in Megabytes Required

So, let's say you have 50 users, this would amount to 5000 Megabytes, which is roughly a little more than 5 Gigabytes. To be on the safe side, let's round this number up to 7 Gigabytes.

But, there are two other factors to consider. This is a requirement for your data only. There needs to be enough room to do data backup, so let's give another 1000 Megabytes to this purpose.

Now we're up to 8 Gigabytes of space.

A rule of thumb is that you should never let a drive get less than 80% full, and you really should have more like 40% of the space free.

40% of 8 gigabytes is 3.2 Gigabytes. Add 3.2 to 8 and you get 11.2 gigabytes.

You can see that even with 50 users, a typical drive of 20 gigabytes is sufficient, and will leave you with a lot of extra room.

OK, so what if your users won't stay within this 100 Megabyte limit? To enforce the limit, space quotas can be implemented on your network. Space quotas hearken back to the old days of mainframe computing and ten megabyte hard disk drives. Because so many users were sharing so little space, a policy-based accounting system for space use must be implemented. Space quotas allowed the system manager to set an artificial upper limit for users' personal drive space, or other shared drive areas in the effort of economizing on disk drive space.

Although such resources can no longer described as "limited", there are still many good reasons to use quotas, like preventing users from downloading the entire contents of their program disks into their personal drive space. There are available many third-party products for enforcing space quotas on whatever network server type you choose to go with. Some server types, especially in the world of Unix, have this quota functionality built-in. The quota software is installed on the server, so you will want to investigate this for the server's operating system that you are using.

There is also the advantage of maintaining shared data space. This can allow your users to work on things as a group, and to save information for future reference. On my network, each user has two data drives, one is the "F:" drive, the other is the "H:" drive. The "F:" drive is private storage space. The "H:" drive is shared. All network users can delete, change, add to, and modify information in the shared directories. We also have several "drop box" type directories within the "H:" drive. These are set up so that certain people can add/delete/modify/see things in the directory. Other users, however, can only add to the directory, and cannot see its contents. These directories are useful for situations where you need to collect data, but do not want to share it with everyone else.

PROGRAM STORAGE

Program storage is similar to data storage. Typically, I run a machine called "tech server" or something like that, which has utilities, drivers, and programs on it. I allow administrative level users write to this network share-point. Regular users can read from it, but cannot add or delete anything.

Tech servers need not have enormous amounts of drive space. I would imagine that a typical hard drive in a desktop computer is ample space for this purpose. It is likely that if you have an unused machine hanging around that has a CD-ROM drive, that you could load a lot of the software onto this machine, and set it up for this purpose.

You will find that by employing a utility share-point, that you can basically store all of your CDs and disks, and refer to them only if you need a backup copy of something.

WEB AND EMAIL SERVICES

Others of these services do not require "servers" local to your site, per se. Electronic mail and web site hosting can be done as easily through a third party as with your own machines.

Why host these services yourself? Chiefly, Control is the best answer to this. Several weeks ago, a local newspaper published an article about the Internet. It was the first cover article in the recent memory of this author, to discuss a technology issue. It distressed this author because it was not a positive article about technology, it was an article about a church website that had come under attack by an intruder.

It seems that someone from the outside was able to change the main web page to read somewhat less than complimentary statements about the Cardinal (the head) of the Baltimore archdiocese. Well, OK, it isn't good that this kind of thing happened, but why is it newsworthy?

Unfortunately for this church, the site was not actually owned or controlled by the organization. It appears that the web site maintenance contractor owned the site. The contractor could not be reached by anyone at the web hosting company, by the church or by the central office of Information Technology. The web hosting company, the people who actually make the website available to the public via the Internet, could not legally do anything to help the Church, because they did not, in a legal sense, own the website. The site remained online with the modified content for several days while people were contacted and the issue was straightened out.

Had the site had been maintained on a local server at the church, or at the very least legally owned by the church, it would be easy to simply disconnect the machine, disable the website, or modify the content back to the way that it was before. The bottom line is that they would probably not have made the front page of the local paper.

This unusual situation aside, there are many reasons to host the website yourself or to have the hosting done by other companies. It is usually more reliable, albeit more expensive, to host with a third-party web hosting contractor. Hosting companies tend to have much better Internet connections than individuals or small organizations. Additionally, they protect their servers to a degree that is usually impossible to do consistently at the level of a small organization.

Backup and virus protection issues on a publicly available server like email or your website are crucial to a degree not found on your other servers. They sit like the outside doors of your network. Curious, outright nosy people try all day long, rattling the doorknobs, trying the windows. They peck around trying to find an opening. If intruders do find a hole, they will exploit the system immediately. It produces headaches for the system manager, and a potentially disastrous situation for the users.

BASE SERVICES

This morning, I walk into my office to a slew of voice mail messages about weird email arriving in everyone's mailboxes. Trouble always hits hardest on Monday morning. I put my normal schedule on the back burner, and look at the email server. The list of messages waiting to be sent is up to almost 1000, almost 100 times the expected amount. I quickly realize that a virus has infiltrated our protection, and has attacked us. Three days are consumed fighting this fire.

October 2002

Although many network managers in the field have their own horror stories, servers that host email and web content can be run relatively safely. Making sure that your virus protection is up-to-date, and securing your servers against outside attack is key. There are many advantages to maintaining your website at your local premise.

The reason that we decided to move our website to our own server was interactivity. We realized that we were capturing only a very small slice of our parish community with our site's content. Our site worked correctly, but the look belayed the fact that our site was becoming obsolete.

We decided that we could use our site to leverage our communications ability to our parishioner base. The primary means of communication between the parish and its parishioners is our bulletin. The bulletin in our parish is the hub of all of the activity therein. It includes advertisements for our upcoming events, blurbs about new happenings, and a schedule of "what's happening" this week. The problem with our bulletin is that it reaches the same group of people every week. When we advertise job opportunities, we catch the same groups time and again. We felt that we could capture a larger segment of our community with alternate means of communication.

I realized that both of these problems were intertwined. On one hand, we were aware that we were not attracting the attention of enough people with our website, while on the other hand we were concerned about involving new people with our existing means of communication.

We started slowly. First, by making our content from the bulletin available online. As time progressed, we moved more content to the website, as well as pictures and other content. This is not to say that we are finished, because websites, like housework, are never really done, and ours has a distance to go.

Depending on your organization, the key advantage to hosting your own site can be control over your backup. Many hosting companies rely on weekly backup. Weekly backup is good for certain sites, but can backfire if your updates are ill timed. Because you have no control over the timing of a system failure, frequent backup of all of your systems is crucial.

Data Backup

Parents protect their children, teachers protect their students. You need to protect your servers. If there were a habeas corpus for the network administrator, it would include backing up your users' data. Backup reliably, change the entire set of tapes on a regular schedule, and by all means, test the backup tapes from time to time.

- Backup Users' Data Regularly

- Run Regular Tests of the backup, by restoring the data to a test location

- Change the entire set of tapes on a scheduled, regular basis

Got it?

The loss of users' data is the worst thing that can happen on your network. Your users will *never again trust your system* if it cannot reliably bring back all of their data when the system fails.

What do you mean, *When my system fails?* Face it, the beautiful new system will one day fail. Given enough time, all systems fail eventually. All mechanical aspects of a computer carry something called a MTBF (mean time between failure) rating. For a hard disk drive, this may be 100,000 hours of continuous use. MTBF ratings are based on lab results from equipment. The equipment lab results are usually garnered in the manufacturers' labs. This does not mean that the manufacturer of your hard disk is dishonest, but all things in the physical realm are prone to failure. If you use the drive for a long enough period of time, it is going to fail you. If your luck runs anything like mine, it will probably happen on a sunny Friday afternoon. Making certain that this is not the last word, literally, is the job of the backup software.

Drive failure is always a likely culprit for system failure, but it is not the only factor in the demise of a server. RAM and motherboard failure poses the potential for loss of information. The difference between RAM and motherboard components and the hard disk lies in the nature of the equipment. Whereas RAM and other components are solid state devices, the various cooling fans and the hard disk drives are the most volatile parts of a computer because they employ mechanical motors to operate. You may notice on older equipment that the fans start to sound a little louder or rougher. The drive is the same way. Because as in the fan, lubricants keep the drive motor working properly, and they, just like motor oil, lose their viscosity over time. Unlike motor oil, you cannot easily replace these lubricants.

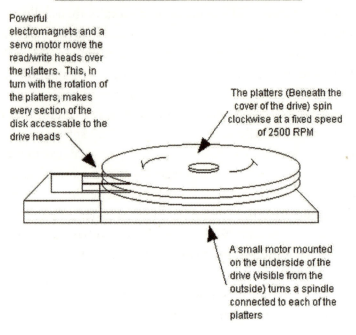

THE OPERATION OF A HARD DISK DRIVE

Powerful electromagnets and a servo motor move the read/write heads over the platters. This, in turn with the rotation of the platters, makes every section of the disk accessable to the drive heads

The platters (Beneath the cover of the drive) spin clockwise at a fixed speed of 2500 RPM

A small motor mounted on the underside of the drive (visible from the outside) turns a spindle connected to each of the platters

FIGURE 2.2: HARD DISK DRIVE CUTAWAY

Your drive may eventually have some kind of motor problem, or may simply not have enough energy to break its inertia and begin to spin up. This happens frequently on older servers that are never shut down. The motor uses a certain amount of power to spin up the circular platters within the drive. If for some reason it cannot spin up the drive, the drive simply will not work when you restart your system.

Additionally, the drive heads hover millimeters over the platters. They use magnetic energy to read and write information onto the platters. If dust or dirt enters the drive, these heads can "crash" into the platters, creating a dead, or unusable space on the platter or in some cases outright failure of the drive.

Are you nervous about drive failure? Don't be. Your servers and drives can obviously be replaced. Any hardware part can be replaced fairly easily, and most server failures do not require that the entire machine be scrapped, anyway. Data, however, cannot be replaced. Data is the only thing on your network that no amount of money or crying can bring back. This is the most important section of this book, because Backup is Important.

Battle this problem with logic. Commit yourself to the idea of a set of rotated backup tapes. I would highly recommend doing a daily full backup every workday of the week.

There are a great number of one-tape and multiple-tape backup schemes. While most backup plans are sound, a few of these schemes seem like they were devised by the pound-foolish, modeled in fits of hysteria over the thought of spending money on tapes. I think that some people would back data up on pieces of toast, if they could devise a way to make it work. We won't discuss the "toast" method here, but we will discuss a few effective but simple methods.

Four Plus Two

"Four plus two" is a sound but simple data backup scheme. It is a standard scheme which is used in many organizations, both small and large. In this scheme, there are four tapes which are labeled Monday through Thursday. The fifth tape is labeled with the title "Friday A". The sixth tape is labeled "Friday B". The first four tapes are rotated for up to six months, and backup the entire data store. The fifth and sixth tapes are rotated and each used once every other week. This tape is taken Off-site, to a secure location.

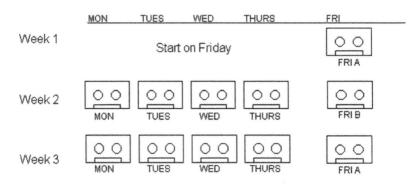

FIGURE 2.3: A TAPE ROTATION SCHEDULE

What is "off-site"? Well, it should be taken (minimally) out of the room, and preferably the building where the server is located. Depending on your feelings of responsibility over the system, this may mean that the tape is taken home with you at the end of the week, or put in a *fireproof* safe. In any case, the tape should be taken away from the building where the server is, for the sake of redundancy. Keeping your eggs in one basket is never a good idea.

On our network, we practice over-protection. We run two completely independent backup systems, each with separate sets of tapes. Our tapes are all DDS-3 tapes, which look like small cassette tapes, and are all disposed of on a six-month basis. There is also a another standard of tape called the "Travan" Tape. Travan and DDS are the two standards for tape backup in a commercial setting. In different settings, I recommend and install both Travan and DDS type tapes. The DDS drives are more expensive, but the DDS media seems to be less expensive. Staying with brand name tapes is a good idea, no matter what tape standard is chosen. On networks that I administer, I require one type of drive or the other, and I only buy the tapes made by the drive manufacturer.

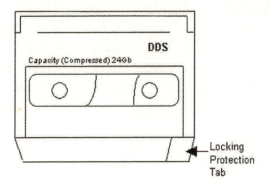

Capacity (Compressed) 24Gb

DDS

Locking Protection Tab

2.4 A DATA TAPE

Data backup tapes are marked with standards, just like audio cassette tapes are marked.

DDS

DDS stands for "Digital Data Storage". These drives can either be type 1, 2, 3 or 4. As the number increases, so does the amount of data which can be stored on a tape. The tape will indicate a number, possibly two numbers.

DDS-3 tapes store 24 Gigabytes of data, if data-compression is used. The tapes must be used with the same, or greater drive. This means that a tape created on a DDS-1 drive can be read by any other DDS compliant drive. A DDS-3 tape, however, can only be read on a DDS-3 or DDS-4 type drive, not on a type 1 or 2 drive. DDS tapes are very reliable, and are readily available from computer supply stores. The strength of the DDS standard is the speed and reliability of the data cassette.

TRAVAN™

Travan is another standard which is popular in the backup arena. This cartridge can store either 20 gigabytes (non-compressed) or 40 gigabytes (compressed). These drives and tapes are sturdy, and reliable, employing a good dust-protecting door. They are proven technology, having been in use for more than a decade to backup computer networks. Travan offers two advantages. Typically, Travan drives are less expensive than other types of data backup drives. The expendable Cartridges, however, are similar or higher in price to other data backup standards.

Our Backup Model

On our system, we backup all of the users' email, email address books, user data, shared data, databases, and the website. We use two banks of tapes, and two tape drives. Additionally, the web and mail server has a redundant tape backup scheme, using the "Four Plus Two" backup method.

There are two banks of tapes, labeled A and B. Each set contains 15 tapes. On the edge of each tape is a typewritten label. The first tape in the set is marked "1–16", the second "2–17", and sequentially on until you get to the 15th tape, which is marked "15–30".

The white adhesive label on each tape looks like this:

FIGURE 2.5: A DATA TAPE LABEL. LABEL INDICATES THAT THIS IS
TAPE "1-16" TO BE USED ON THE FIRST AND 16TH OF THE
MONTH. IT IS PART OF TAPE SET A, AND WAS PLACED IN SERVICE
ON FEBRUARY OF 2002

The tape is used only on the dates which correspond to the numbers on the tape. For instance, on the 1st and 16th of the month, tapes from set A and B marked "1-16" are placed in their respective tape drives.

There are two machines that do nothing but run the backup software. These computers both are programmed to run a full backup of all data on the network, onto these two sets of identical tapes. Why two sets of tapes, you ask? Sometimes, there is a problem with at one of the tapes in the set, this ensures twice the chance of a good backup set. Tapes are not the most reliable medium, but they are not perfect, and they do have problems from time to time. It is a very good idea to run two sets of backups.

On months with the 31st, no daily backup is run, and thus no additional 31 tape is necessary.

Additionally, there is a set of end-of-month tapes. These tapes are labeled "March 2002, Set B", et cetera. There is a new tape made for every month. This tape is substituted for the regular tape and is run on the last calendar day of the month that our offices are open. It is the same backup, but is an archival copy of the network data. For safety purposes, this tape is kept outside of the system room, across the street. This ensures that a (fairly) recent backup exists in the event of a massive failure, or if a disaster such as a fire occurs in the building or the system room.

If for nothing else, backup can relieve users, allowing them to put additional trust in your system. You may wish to advertise your data backup strategy to your users, showing them how the tape backup scheme is set up. Advertising the data backup strategy is a great way to get people to put their data on the network, as opposed to on their local computer or floppy disks, both of which are more prone to failure than your tapes or servers.

I have found that little good comes of having information stored on a user's computer. Users' computers tend to have many more extraneous programs running on them, and more problems with data loss and hardware failure. In fact, Failure rates are almost twice as high than on their server counterparts. Servers, possibly because they are more of a "clean room" application, are much safer alternatives.

Data Backup Documentation

People get sick. They take their two weeks vacation, they may even take their two weeks and quit! Things come up in life which prevent the tapes from changing. A day or two of no tape changes is probably OK. Tape drives are actually designed to run the same tape for 5 days or so, because some people don't have a tape for each day.

A data backup scheme will get really fouled-up if one goes for several days with the same tape. This confusion will be compounded should the need arise to restore a backup which is not the directly previous backup which was made on the system. This is because the numbers on the tapes are no longer coordinated with the dates on which the backups were done.

To counteract this problem at our organization by writing a scripted instruction sheet, which allows a non-technical "Trustee" of the system to do regular tape maintenance in my absence. I have included a sample sheet in the appendix for your review. While the technical details of this document are unimportant for this discussion, you may find the format to be helpful. After you have written your instructions, do a few test runs.

Follow the instructions to the letter, as if you had not written them, because the network trustee will do the same. Additionally, describe what equipment and tapes look like, and where they are located. Give the trustee some basic trouble shooting information. In writing a document like this the tendency may be to give the trustee lots of background or extraneous information. It is more helpful to give only what they need to complete the job at hand.

Your tape backup plan should be documented, with certain information about the nature of the backup process.

This document should include:

• A clear title, and the date of the revision of the document

- A brief explanation of what the backup's goals are, and how it works, listing the software package controlling the backup

- A section describing the tape rotation methodology used in your backup scheme

- A section describing the times that the backup runs

- An up-to-date explanation of what is being backed up

- A section listing the types of tapes used, and the replacement schedule for the tapes. This should include the coding scheme used to determine when the tapes were put in place

- A section regarding regularly scheduled full backups, for archival purposes

- A section regarding the handling and security of tapes, including the practices of the destruction or archiving of old tapes, for security purposes

- A section regarding scheduled testing of the backup tapes and system

- A section regarding scheduled regular cleaning of the tape backup drive(s)

A secondary document should be written describing how exactly to run the tape rotation schedule, and how to run the restore function of the tape backup software.

These documents should be dated, and kept up-to-date to reflect changes in the servers or network backup scheme.

A Word on the Care of Tape Drives

- Always use brand name tapes

- Always use brand name cleaning tapes

The tape drive should be cleaned at least once per week, initially using the cleaning cassette that was packaged with the drive. When all of the tape in the cleaning cartridge is used up, replace the cleaning tape with the same brand of tape. Usually, the manufacturer of the drive also manufactures tapes and cleaning tapes. Tape drives are finicky and expensive, and irregular cleaning or failure to clean the drive can cause irreparable data, or data loss. There is very little else in the way of regular maintenance necessary on a tape backup drive. Occasionally, depending on the drive manufacturer, there may be regular maintenance items which must be performed. A typical maintenance item may be "drive alignment", where a test tape is used to test the alignment of the read/write heads, and adjust them if necessary. I do the alignment test every six months when I change the tape set.

Regular cleaning and drive alignment are two proactive strategies that can work to ensure greater life expectancy and reliability of the tape drive.

CD-ROM and DVD-ROM?

CD-ROM and DVD-ROM are the most recent generation of a long family of optical storage products for computers. CD-ROM writers are fairly commonplace now, and are a cheaper way to backup data than using tape, both in terms of the cost of the drive and the media (the discs). The CD writer may cost hundreds, versus thousands of dollars for a tape drive. The chief problem with optical disks is that, well, size does matter. It is likely that they are simply not sufficient in size to back up all of your users' data. There is a theoretical maximum of 700 Megabytes of data that can be fit onto a compact disc. Although much greater in capacity than recordable CD, rewritable DVD is currently an expensive option. As time goes on, it will no doubt become cheaper and will likely supplant CD-ROM completely.

Capacity limitations notwithstanding, disc may not a bad option for smaller networks. Typically, you will want to backup your data in a manner that allows 1 disc/tape/etc. per backup session. If you use CD-

R or CD-RW, you will likely be faced with switching the disc halfway through the process. This negates the possibility of running your backups at night. The advantage of nightly backup is that reduction in network speed performance as a result of the backup software will make no difference to the users of the network. CD-R and CD-RW do not have the capacity to back up even small amounts of information effectively. As compared to tape, which can backup 40 or 50 Gigabytes, even DVD-R and DVD-ROM only backup in the neighborhood of 4 1/2 Gigabytes of data. 4 Gigabytes may be more than you have today, but will likely not meet your needs in the near future.

My problem with compact disc as a backup medium is its inherent unreliability. Compact Discs (and by extension, DVDs) take very little abuse before they cease to function. Tape continues to work, even if part of the tape is damaged. Tape, in general, is much more impervious to damage than optical media. Although the technology continues to improve, I have personally had numerous discs that stopped working for no apparent reason whatsoever.

If you do choose to go in this direction, bear in mind that some data backup programs will not work with CD or DVD drives. You will want to check out the backup options before making a purchase decision.

Centralized Virus Protection

In the course of network security, you are bound to run into the question of virus protection. Whether this question is posed by your staff (very possible), a flyer in the mail (more likely), or by an anti-virus salesman (just kidding), it will become an issue for you. There are a number of good options for antiviral software, most of which are readily available. Put it into your plan, because the question is not if, but which package you will need to go with.

There are two courses of action with antiviral software. The first, simpler one is to install antiviral software on each of the client and server machines, and to either update each yourself, or have the clients update their software individually. The alternative to this is to buy a different version of the antiviral software, which is designed for centralized systems. I have found that the best remedy for this is to centralize, rather than to allow the users to do the updates, which almost invariably they will not do. If you made a list of the top four antiviral packages right now, each of them has a facility (usually in the form of an "enterprise version" of their software) which allows you to manage the network as a whole.

The centralized approach is, up front, more expensive to purchase, as it usually requires many licenses and some additional software. Before the price difference makes your decision, think of the problem this way. If you have a virus infection, it will take your network down for at least a day or two. How much is this time worth when you figure in all of your users' time value? If you calculate it this way, the decision may become easier to make.

Centralized Network Time

Many network operating systems have a provision for a "time server". The time server is either an internal function set by one of the servers' own clocks, or by an external (Internet-based) clock. This is dependent on the particular time server package that you opt to use. Typically, the time server that is included with the network operating system will use the clock in the server that it is running on, rather than from an Internet-based third party.

Time servers can be made to run as a command in a logon script, resolving the date and time of the user's computer with that of the network server that controls the time. This makes certain functions, such as backup (if you choose to do backup of a client's PC), virus updates, and other time-dependent functions run correctly. It also has the soothing effect of coordinating the time on all the computers, which has the appearance of being much more professional. Sometimes, perfection *is* in the details.

Centralized Login Scripts

Centralized logon scripting is a feature which is available on most all modern networking software. This functionality allows you to attach drives and printers, set the attributes of a users' computer, and even update or install programs on login. It only works as long as the users log in and out on a regular basis.

Centralized Printing

One service that should be added immediately is centralized printing. Centralized printing allows one network server to act as a "print server", making the printers available to users on the network. This allows the network manager to maintain a log of what has been printed, and when. It also allows for prioritization of print jobs.

The key advantage, though, to network printing is the ability to control (1) which printers people can print to, and (2) how many copies of something that they can print at once. This allows you to prevent people from using color or specialized printers. It also allows you to prevent people from mistaking the word "printer" for the word "photocopier". Laser type, and especially ink-jet and impact (dot-matrix) type printers are not built for production in the way that most photocopiers are. Allowing users to print a certain limited number of copies prevents the regular use of these machines for duplicative purposes.

THE PHYSICAL INFRASTRUCTURE

At this point, some diagramming may be helpful to determine the plan for your network. You are moving from the ethereal realm of theory to the physical realm of hardware and wires.

Make a diagram with high-level services on the left hand side, and servers on the right hand side. This should allow you to design the basic definition of your network server-side equipment.

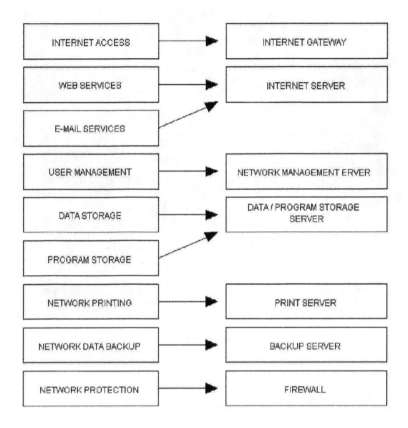

FIGURE 2.6: NETWORK SERVICES DIAGRAM

Once you have your diagram of servers and services, it will become easier to visualize many other aspects of the system. Some of the most basic networks may need as few as one or two servers.

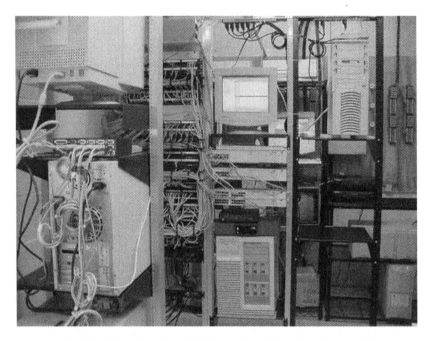

FIGURE 2.7: A SERVER ROOM SHOWING WIRE MANAGEMENT
EQUIPMENT, SERVERS (CENTER) AND RELAY RACKS

NETWORK DESIGN

Building a network is a lot like building a house. When you build a home, you need to specify different service level items, like Kitchens, Bathrooms, and Bedrooms. Rarely will people design homes with no regard to the purpose of the space.

There are considerations at different levels when building a home. At one level, the designer is thinking about sizes and shapes of rooms, locations and aesthetics, and cost factors. At another level, mechanics and engineering issues are being considered in all different areas, including heating, electricity and plumbing—quarter inch over four feet, and all that stuff. At still another level, hardware and issues concerning building materials are being considered and discussed. All of these levels intertwine, and none are "more" or "less" important than others. If talking about an engineering issue, you are also talking about Aesthetics, and cost issues as well. This balance must be observed and maintained, lest the end product (The house) will falter.

Computer systems generally start out differently than homes. Homes in America start out as a plan, and are built, inspected, and delivered complete to the buyer. The buyer moves in, he assumes responsibility over maintenance over his house, and he uses the services that it provides: Shelter, warmth, a place to store all his Barry White albums.

Conversely, all too often computer systems start piecemeal. Like if a garage, a bathroom, and a kitchen just popped up out of nowhere in the middle of a field. Sounds like the beginning of a bad joke, doesn't it?

Somewhere along the line, people started putting a computer here or there, maybe a printer. There is no network between these machines, no servers or intercommunication. It is likely that other than disks, there is no way to transfer information between users.

FIGURE 2.8 NON-NETWORKED COMPUTER SCHEME. NO ABILITY
TO CONNECT BETWEEN EQUIPMENT OR SHARE PRINTERS EXISTS

You may be thinking that it would be nice to make this system grow. The next logical step is to install network cards in these computers, and connect them together. The most common operating systems allow you to set up simple share-points, which can be accessed by all of the other users. Notice that there is no centralized server. Because this is a peer-to-peer network, each computer is responsible its own information. If one computer is turned off, its information is no longer available to the rest of the system.

FIGURE 2.9: NETWORKED COMPUTER SCHEME. NOTICE THAT ALL
MACHINES CAN SHARE NETWORKED PRINTING EQUIPMENT.

For many offices, this peer-to-peer arrangement works very well. There is nothing inherently wrong with the peer-to-peer network. Administration of peer-to-peer systems is very simple when a limited number of machines are involved. For these types of systems, backup and antiviral protection should still be a high priority. These issues are usually easy to handle with off-the-shelf software packages. For a three to ten machine network, a server is usually more a nuisance than anything else. What does this mean to you? Well, the best advice that I can give is to stick with standard types of networks and network operating systems, so that you can quickly add services if the need should arise.

In the case of a larger system, maybe more than ten machines, a server is usually more-or-less required. There is an inherent problem with peer-to-peer networks. Networks are normally set up with user names or passwords to control access to information. In a three-machine network, setting up passwords for each share-point may not be necessary. Passwords would be a simple task to implement, and might be needed if there was sensitive information. As the system grows, so grow the number of passwords. Many peer-to-peer networks require usernames and passwords for *each machine,* as opposed to one centralized usernames and passwords that would work on all of the machines. These kinds of systems are notoriously hard to manage, because a roster of users has to be maintained for each separate computer on the network.

One direction that a growing network might take is the one-or two-server scenario may include a small organization who will not maintain local email or a website. This type of network might look like the following:

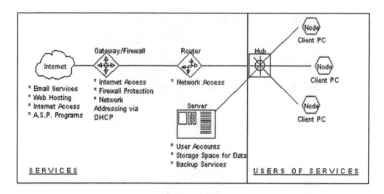

FIGURE 2.10: A SAMPLE NETWORK SHOWING POINTS OF SERVICE
AND USERS

The **server** handles all administrative functions, data storage, and has a tape backup drive. It can also be expanded to handle group printing, fax, etc.

FIGURE 2.11: A TYPICAL SERVER WITH MULTIPLE DRIVES

The **hub** is central to the network, and allows all of this equipment to be physically connected together. Your hub does nothing in terms of network administration, and simply acts as an electrical outlet does. You plug everything into the hub, and everything sees everything else.

FIGURE 2.12: HUBS (WHITE) AND PATCH PANEL, SHOWING PORT
NUMBERS

The **Firewall,** in this particular arrangement, also has a "DHCP" server. While not getting overly technical, DHCP is important because it allows all of your machines that are connected to it via the **Hub,** to have network addresses in the case of a TCP/IP type network. Network addressing, which is essential to the operation of a network, is handled by the firewall. For a small network (under 50 machines total), it is the best option. Using a Firewall with DHCP allows you to have the mechanics of the network working, while issues such as addressing of the individual computers and routing for the Internet connection work without you having to manage issues surrounding addressing.

Another direction for you may be to provide Internet services and have no servers. It is very possible that your organization has no need for centralized file sharing, but would like to have centralized Internet sharing or email accounts. It is then very simple to add things like shared printers, servers, and other devices without messing up this basic structure.

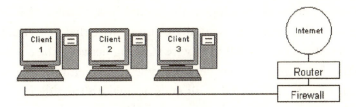

FIGURE 2.13: NETWORK DIAGRAM WITH INTERNET ACCESS
THROUGH ROUTER

With time and advancements in technology, your network will change. For instance, when we began, our network's connection to the Internet was very simple. As I write this book, my Internet connection is in its third iteration, having changed to accommodate first a high-speed connection and firewall, and more recently a local web and email server. To give you a more detailed idea of what your network presence plan might look like, here is a basic plan for a network with Internet access and a local web server. Notice that it accommodates access from the both external Internet clients and from the internal network.

NETWORK FUNDAMENTALS

Most networks today start out as systems of copper wire. As your network grows, as does the complexity of the web of copper wire running through your building.

Although possibly not practical for every network administrator, It is a good idea to be aware of the physical realm of your system. Networks are first and foremost webs of wire. This wire snakes all around the insides of your building, right beneath the skins of ceiling, carpet, and painted drywall. You don't see this stuff, but it does exist, and it rears its head from time to time.

Networks are classified by the manner in which the wire is run. A few of the original network schemes for personal computers employed something called a "Bus" topology. A *topology* is a logical roadmap of the network. Just as a "clover" pattern in a highway defines the pattern of the roads, network topologies are named in a manner of how they are laid out. Bus networks created a chain of wire that was attached to each computer, creating a line of wire running between the computers. Alternatively, there are other network types which employed a continuous line of wire, connected at all ends, to make a roughly figured circle. These types of networks are called Ring networks. Below is an illustration of these two network topologies. In the diagrams below, each numbered circle represents a client computer, file server, or printer in a network.

Neither of these topologies is particularly popular any more in anything but older network schemes, and it is likely that you will never encounter them. Most physical networks today basically follow a third type, which is called the "star" topology. Why are these other models being described here? Bus and Ring networks are a proven, viable model for networking. In many circumstances, a bus or ring type network may be suitable for your applications. In a very small network that does not span more than one level, or group of offices, Bus and Ring networks can vastly simplify networking. Depending on your office space, you may be able to run cables between desks, non-invasively. This cuts down on the amount of wire necessary to complete a working installation, and by extension, the cost of implementing your network.

There is one negative aspect of both ring and bus topologies. Networks of Bus types in reality are a chain of computers with wires running into and out of each network card. Unlike most star networks, which use wire called STP (Shielded Twisted Pair), which looks like thick telephone cable, Bus networks use coaxial (cable TV) cable. Cable TV wire is notoriously difficult to work with, much more so than STP cable. In the short term, the Coaxial cable (Called Thin-

Net), is more expensive to purchase, although it is ultimately cheaper because it eliminates the need for a hub, which can be an expensive option.

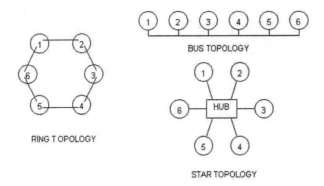

FIGURE 2.14: THE THREE BASIC NETWORK TOPOLOGIES

The predominant type of network today is a star network. Actually, it is one particular kind of Star network, which we will talk about later. A star topology network assumes a Hub in the logical middle of the network. Each client, server, or other network device has a piece of wire running from itself exclusively to the hub.

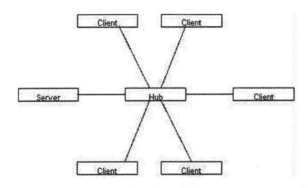

FIGURE 2.15: THE MOST COMMON NETWORK TOPOLOGY: THE STAR NETWORK

The hub acts as the center of the star, and the wiring runs between (simplistically) the hub and the computer.

Our network is set up in a star topology. Like many networks, it is more complex in design than a simple star. The server room in the main building acts as a hub of the star. The other buildings act as legs off of this hub, each with their own wire closet and their own star topology within that building.

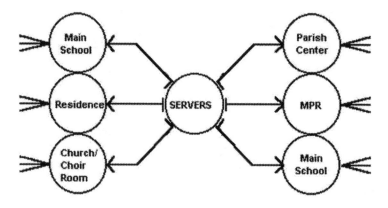

FIGURE 2.16: A META STAR NETWORK

In the above diagram, you can see the pattern emerge. The server room is the logical center of the system. The inside legs of this system are fiber optic cables. We have six separate buildings on the campus which are interconnected: Two School Buildings, a Parish Center which contains offices, a Multipurpose Building (MPR), A Residence, and the Church building. The plant also has a maintenance building and a house, but as these two buildings are not connected to the network, they are not pertinent to our network plan, and are not listed as such. In your network topography, you may choose to include all of your structures, or just the ones which are connected to the system.

Each client computer is connected to a piece of wire which connects to the jack in the wall. The jack is numbered on the wall, and again in the wire closet, where there is a patch panel. A patch panel looks like a new version of the old telephone operators' switchboard. The back of the patch panel has all kinds of wires coming out of it, each going into the ceiling, floor, or wall through large tubes, called Conduit. These wires go to the jacks in the walls for telephones or network equipment.

On the front of the patch panel, each jack is labeled with a number which corresponds to the jack. The patch panel serves as an intermediary demarcation between the network wire and the hubs. The network hubs have smaller wires, called patch cables, which connect to the patch panel.

The most popular, and standard type of star network is called a 100BaseT TCP/IP network using RJ45 connectors. For most installations, this is the type of network which is preferable to use, because it is simple enough to be maintainable, but scalable enough to be able to accommodate growth. What does this mean to you?

WIRING

Network wiring has evolved significantly over time. Originally, network wires were thick, thicker than cable-TV wire. This wire was called "ThickNet", or sometimes "Frozen Yellow Garden Hose", because it was rigid, (usually) yellow in color, and the approximate thickness of a residential water hose. Large adapters, appropriately named "vampire taps" clamped over a length of the wire, their prongs attaching to the metal conductor embedded within the casing. The envisioned network plan was that the wire and adapters would reside either in the plenum above a drop ceiling, or beneath a false floor. ThickNet wire was basically too thick and rigid to allow it to be fed down walls, so the wire between the adapter and the computer had to be run down the wall, or up through the floor. This was an unwieldy arrangement, but for the acceptable standards of the time, it worked well.

As time progressed, thinner coaxial wire (of the television cable variety) replaced the Thick Net standard. ThickNet still had an advantage of speed-over-distance, because of certain physical superiority and immunity to something called "loss" which is the loss of signal over a distance of wire. Thin-net did not require the extra adapter, which was expensive, and it was thin enough to reasonably run through surface mounted wire management equipment (wiremold®), or through walls. This was a major boon to network equipment manufacturers, as it made installing networks a lot easier.

In parallel to this, Apple Computer was implementing a networking scheme for Macintosh computers. Apple's network used a multi-pin mini-DIN adapter, which they were trying to use for everything on the Mac. Apple's network was a bus topology, and required that each computer have an adapter connected to the Mac's "Printer" port, an RS-422 (faster) serial port. The adapter had a short wire connecting to the Mac, and two ports, one into the Mac from the previous device on the network, and one Out for connection to the next device. Farallon, a firm who made Apple peripheral products, devised a way to use standard UTP cable (Unshielded telephone wire) to replicate the networking capability of Apple's $60 proprietary network interconnection cables. This in and of itself was not remarkable. Lots of firms were replicating Apple's proprietary hardware with less expensive versions. The real benefit of this system was that it used the *alternate pair* in the phone cable. What this means is this: In a telephone system, there are always 2 cable pairs. 1 pair (two wires) are used for the telephone line itself. The other two wires can be used for a second telephone line. If there is not a second line, these wires are not used. The brilliance of Farallon's system was that it allowed your *existing* telephone wire and jacks in your walls to extend the network throughout your home or office. Many initial networking schemes were based on this simple, inexpensive system of wiring. It was by no means fast, but it worked remarkably well.

RJ45 connections, shielded twisted pair, and the 568b standard

In time, as will happen, people coagulate to a standard. In the case of networking, people have basically standardized on the following network:

- It is a STP Cat5 or Cat5e (Shielded cable with4 twisted pairs of wires inside)

- It employs RJ45 Connectors and Jacks which are labeled in such a way as to indicate that they comply to Cat5 or Cat5e standards

- The jacks and patch panels comply to the EIA/TIA 11801 standard T568B wire map standards. What this means is that each the eight wires each cable are connected in a standard way to jacks and patch panels, so that everything works together correctly.

Physical Attributes of a 100BaseT Network:

100BaseT networks are a standard type, which can include a variety of types of supported network devices. Devices may be laser printers, desktop computers, printer sharing equipment, and servers, all of which are connected to each other via wire in a star topology.

100BaseT is an industry standard defined by an international, vendor neutral industry association. "vendor neutral" means that this is a presumably objective group, which is not trying to market a particular brand of products. Please note that these labels are not brands, but names of standard types of products. Most 100BaseT networks conform to the Ethernet standard. Ethernet is a trade name based on the electrical transmission specification, one that is used for the computers to communicate to each other.

100: The discrete components of the infrastructure are certified to operate at a minimum speed of 100 Megabits per second transfer rate between any two devices on the network. Please note that many computers will not achieve this 100 megabit transfer rate, because they use older, 10 megabit maximum Ethernet adapter cards. Just like many modems cannot achieve the maximum speed, many network cards cannot in reality achieve their maximum speed ratings.

<u>Base</u>: All Ethernet networks operate on a Baseband principle. Only one device can communicate ("Talk") at a time. When you talk on the telephone, you must wait for the other person to stop talking before you can talk. Think of an Ethernet network as a continuously operating conference call with dozens or hundreds of different parties online.

<u>T</u>: This indicates that the network uses Category 5 (a quality assurance rating based on the quality of the copper and which rates speed), Shielded Twisted Pair (STP) wiring, similar in physical structure to telephone wire, but with 8 wires instead of 4 or 6. Although Ethernet typically uses STP cable, some smaller or older Ethernet networks use the aforementioned coaxial (cable TV) wire, which is the other standard called 10Base2. 10Base2 networks cannot operate at as high of speeds as STP networks, because of the nature of the coaxial wire. This type of network can operate at 10 megabits per second, roughly 10% of the speed of the 100BaseT-type network.

<u>RJ-45:</u> Because the network uses twisted pair wiring, it must employ a connector which can be attached to it (is compatible). The only type of connector which is satisfactory is a type called RJ-45. A normal telephone uses an adapter called RJ-11 (the clear plastic plug), which has a maximum of 4 wires. RJ-45 looks very similar, but is bigger, and has 8 wires instead of telephone's 4 or 6.

Incidentally, network wire is usually referred to in terms of how many pairs of wires are present in a cable. An eight-wire cable is referred to as a "4-pair" cable. Four-wire telephone wire is usually referred to as being "2-pair" cable.

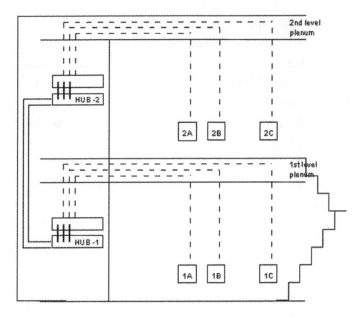

FIGURE 2.17: A MULTI-LEVEL BUILDING WITH RISER

Most physical networks follow the above fundamental structure. In most new construction (offices and other commercial sites built within the last 10 or 15 years), there are several features which make installation easier for the installers and maintenance people. These features are dropped ceilings, studded drywall, and metal or PVC conduit.

Drop ceilings are a type of acoustical tile material which are manufactured in rectangular blocks, and sit in metal tracks which are attached using heavy gauge wire to the actual ceiling. The actual ceiling is made of dock planks, or cement sections which run the length of the building. If done correctly, there should be a 2 or 2 1/2 foot gap between the drop ceiling and the real ceiling. This is referred to as the *plenum*. Your network wire is not alone up there, and shares this space with a host of other things, such as water piping, electrical lines, and HVAC (heating, cooling and ventilation equipment) ductwork.

The plenum is where all of the wire for your network runs between halls and rooms, and ultimately to the wall where it is to be installed. It is important to note that code requirements call for a specific type of fire retardant wire, called "plenum rated" network wire. Plenum rated wire is more costly than non-plenum, but its advantage is that, if burning, its "plastic coated" shell will not produce a noxious fume. Since ventilation in a modern building is linked inexorably to the plenum, this feature of the wire can be the difference between life and death in the event of a fire.

The wire will ultimately have to travel down the walls to the data jack. Most offices employ some form of pre-fabricated drywall. Modern drywall usually has a certain amount of space between the outside wall and the back of the drywall, or between two pieces of drywall. This amount of space varies depending on the installation. If the wall is an inside wall using two-by-fours for structural support, typically there is a two-inch gap between the drywall panels.

Most building codes call for a stud (a wooden or metal beam) to be installed predictably every sixteen inches. When installing data wire, it is easier to drop the wire between these studs, and into a box mounted within the drywall, which houses the connector, and from which the visible faceplate on the outside of the wall is attached. A stud-finder, an electronic device that is readily available, should be used to determine where the beams are located within the wall. It is also important, in the case of an existing construction, to check with local electrical codes. Although at the time of this writing, no licensure is required to install network wiring, there are codified requirements regarding distance from high-voltage electrical wire and equipment. Likewise, there are reciprocal requirements in data networking standards regarding distances from electrical service wiring.

What if you have plaster walls, or some other form of non-drywall installation? Typically, the best option in this type of situation is a surface mounted option. Many companies sell surface-mounted systems, complete with ceiling and wall installation hardware, and accessory boxes for installation of network jacks.

Finally, in a multi-floor building, or multi-building installation, there must be a series of pipes used to direct the network wiring in the ceiling back to a centrally located punch-down panel. These pipes are called *Conduit*, and are made of either metal pipe or a plastic ABS or PVC pipe. Usually, these pipes are installed during the fabrication of the building. Conduit is crucial, chiefly because it protects the network wire from damage. Damage to wire located between buildings or floors of a building is virtually beyond repair, in most circumstances.

In the preceding figure, the dashed lines represent copper wire, of a type that is suitable for network usage. Network wire is literally twisted, and is different than the quality of wire necessary for most telephone system use. The darker lines represent short interconnection cables of the same type that is run in the walls and ceilings. This interconnection cable connects each piece of wire from the jacks into the hubs. This allows hubs and equipment on peoples' desks to be connected together. Finally, in a larger installation, there is a "riser" which connects two or more floors of wire together. Riser cable may be made out of Cat5 or Cat5e certified copper cable, like that in the walls and ceilings. In larger installations, it may be something faster than copper wire, such as fiber optic cabling. Depending on variables such as your building type, the speed required, and the amount of money you wish to spend, your network may differ from that of this plan.

Although you may never have the opportunity or need to install any wire in your own building, it is good to be aware of the method in which it should be done. When installing network wiring in a drop ceiling, try to be courteous to others who share this space. Always place your wire as high and out of the way as is reasonably possible. Try to stay out of the way of other major in-ceiling applications, like ductwork. Aside from simple courtesy to others who use this space, this will protect your wire against other service persons who maintain the other electrical and HVAC equipment. I have found that "low voltage" network and telephone wire is open season amongst some of these guys.

You will want to supervise the installation of network wiring when it occurs in your plant.

Before you call the network people, visualize where the jacks should be located in your building. All too often, data jacks are not located near where desks or other furniture is likely to be placed. If you do not make the decisions based on logical placement, the network contractor will make the decision based on convenience and ease of installation.

You may also wish to pop your head into the plenum prior to calling the contractor. The area above the drop ceiling is important to be aware of, because there may be major issues to consider. Sometimes, you can see these issues as well as the estimator. If there is a great deal of ductwork, or a solid cement wall impeding the run of the cable in a building, it may increase the cost of the installation.

The summers at St. Joseph's are always hectic for the staff. There are three months in which to bring all of the repair issues back into check. As with all schools, major improvements are often done during the summer months, when the buildings are open and available for work crews. Last summer, we decided to replace the aging Public Address system. The PA system had been installed over decades at the school, and some of the equipment was beginning to show its age. It was a tough decision, because we had just had a major wiring repair completed on the old system, at significant cost. We decided, though, that the extra money involved in repairing the old system would not be recouped, because we would ultimately need to do the repair anyway.

Without suffering through the details, we went through the needs and RFQ process, and wrote a proposal and request for quote. In due time, the bids came, and we made a decision, with the same company whom had done a large sound installation for us in recent past. The system was ordered, and one sunny morning three men came to install the new speakers, and console.

They had assumed that because the wiring was new, that it would be adequate for our new system. Unfortunately, the wire used was not the correct type for the new system. Each classroom had an individual wire running from the console to the speaker in that room, even if that room was in a different building. The individual wires running between the three school buildings were two conductors each, meaning that each cord contained two wires inside. The contractor was quite nervous, because they did not anticipate adding new wire as part of the installation, a situation which would have cost them money to do our job.

This was an oversight on the part of their estimators, in not calling the company who manufactured the new equipment to see how many conductors each speaker would need to have. The terms of the contract would have called for them to pull a great deal of wire as part of this job, but at a significant additional cost to our organization, and at a significant time increase to completion. If we had pulled wire, we would probably have been working on this system well into the school year. As there were very few satisfactory ways to get new wire between the buildings, we worked together to find a suitable and workable resolution. We were able to find a resolution through using the shield of each of the existing wires as a third conductor for the speaker. The sound quality was not affected in a noticeable way, so it worked for us, and the sound contractor did not have to charge us a great deal more money. Additionally, the project stayed on schedule, losing only one or two days overall.

While this was an example of a cooperative compromise, you want to keep an eye on your contractors to make sure that they are not short-cutting your job. Because you probably do not work in the field of wire installation, the only suitable way to ensure quality is through certification and standards adherence. Reliable standards are created by vendor neutral organizations to ensure proper installation practices and materials use. Enforcing standards and certification help you to ensure that your contractor has hired professional, experienced people to install your wire. They also help to show you that your contractor has purchased quality, brand-name materials that were manufactured to certain quality standards set out in the specifications. It is a good idea to make sure that the network contractor (or you, if you installed the wire) provide two things. First, each jack should be clearly numbered with a logical system of numbering. The IEEE (The International board who defines networking standards and practices) has defined standard procedures for numbering and coding data jacks. If you choose to install your own network, It is a good idea to learn and follow these practices.

Second, the network should be documented. At the most basic level, drawings of the wiring schema should be provided, in the form of blueprints showing the locations of data jacks and key premise equipment, such as wiring closets and patch panels. Additionally, a videotape or written description of the locations of wire should be requested of the network contractor, and kept in a save place for future reference. This is especially helpful in a new construction or building renovation projects, where walls and ceilings may be covered up during the process, and may not be as readily accessible as they were prior to completion.

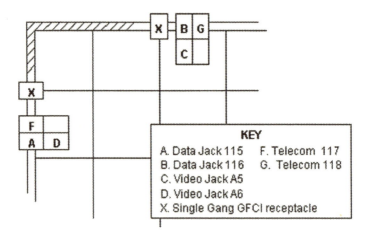

FIGURE 2.18: PARTIAL FLOOR PLAN SHOWING OUTLETS

TCP/IP: THE LOGICAL SIDE OF A NETWORK

Because of the Baseband nature of Ethernet, there must be a methodology in which communications occur on our network. If every machine talked at once, it would be constant pandemonium, and nothing would ever work correctly.

Because of this, there is a communications standard called TCP/IP. TCP/IP is two discrete rule-based components, combined to create a viable internetworking system.

TCP: Transmission Control Protocol (TCP) allows computers to detect that other computers are using the network. A computer first "listens" to the network.

If someone else is talking, it waits awhile, and checks again later.

On the other hand, If there is no one is talking, the computer begins to "talk" to another other computer or printer. All of this talking is how YOU check email, print a file, save, or open anything. There is also a lot of maintenance talking that goes on, behind the scenes.

Sometimes, literally between the "listening" part and the "talking" part, someone else will begin to talk. This causes a collision. Collisions happen many thousands of times per second, and are normal. TCP makes sure that the collisions do not adversely affect the operation of the network.

IP:

Just like your home has an address, each computer on any network must have a unique address. In many ways, network addresses are like telephone numbers. On TCP/IP networks, there are even special servers that act like phone directories, automatically matching up the easier-to-remember network names with network numbers.

Each of these devices has a unique address. Sometimes, the network manager devises an assignment for each address, other times it is done automatically.

This is how the computers on all TCP-IP networks talk to each other. For instance, my computer's number is 10.0.0.161, and the network file server's number is 10.0.0.2. When I use Word to open a file on the server's hard disk called "Memo to Bill.Doc", My computer listens to the network, if it is not busy, it says "hey, 10.0.0.2, I need that file called "Memo to Bill.Doc"). If there is a collision, it waits and tries again. "waiting" means less than a millisecond. It happens so fast that you don't realize it's waiting.

The important thing is that there can never be duplicate network addresses on the same network. If there are two computers with the address 10.0.0.161, then the computers cannot communicate. This is a very serious problem, and in a Static network comes back to one thing: Organization. If the network manager is careful, and keeps track of the addresses, there will be no chance of having this problem.

To ensure organization and prevent IP number conflicts, network managers create lists or databases, called "IP numbers" lists. They reference the list when adding a new number to make sure that it is not a duplicate.

Whether using a Static or Dynamic (Explained later in this chapter) networking scheme, as a network administrator, you should always know, or be able to find out a computers' network address. Knowing the IP address of a computer can significantly aid the troubleshooting of a multitude of problems on the network.

TCP/IP was invented almost 30 years ago by the army and select higher educational institutions for use with the Internet. The Internet is a large meta-network (a network of networks), and uses this protocol exclusively. The fact that the Internet is a single mammoth TCP/IP network should illustrate the scalability and reliability of the protocol.

The rules and methodology surrounding TCP/IP can expand into a library of books. If you are interested in the workings of TCP/IP, there are many books out there to guide you. Later in this section, a different kind of TCP/IP scheme will be described which may work well for your small network.

For you, the important thing about TCP/IP may be that it is a numbering scheme for computers. There is a lot of logic, mathematics and lots of rules behind TCP/IP addressing, most of which are not important at this point in time.

It bears saying that TCP/IP is one, but certainly not the only network protocol. In this time period, it is really the only protocol that should be seriously considered when planning a new network, especially if Internet connectivity is desirable. There are a host of other network protocols, some of which are simpler to configure and implement. However, TCP/IP is a good, very stable protocol which allows for unparalleled scalability, and routing capabilities which are two key issues, even on a small network of computers.

The key advantage, and the real reason why other standards will not be discussed at length is that TCP/IP has become the standard of network protocols. If you are looking at buying new network interface equipment, you will be hard pressed to find hubs, routers, or network cards designed for systems other than those described in this book. If you deal predominantly with donations or older network installations, you may be an exception to this rule.

Static and Dynamic IP Number Addressing

There are generally two types of TCP/IP Addressing schemes: Static and Dynamic. Static addressing means that you assign each address to each new piece of equipment that you add to the network. Dynamic means that a piece of equipment running a DHCP server attached to the network handles the addressing. More and more networks are using DHCP, which is much easier to maintain than static addressing schemes.

On my network, I use a Static-addressing scheme. Despite the overhead and additional maintenance that this requires, static addressing affords several distinct advantages. First, you can control each new machine, as the holder of the master list of available addresses. You become the "phone company", in essence. Especially if you are in an environment as we are, where there are a lot of available data jacks everywhere, you can know that the only machines participating on the network are likely to be ones that you placed there.

Secondly, DHCP is what is referred to as a "chatty" protocol. There is a lot of overhead traffic associated with DHCP that is of no direct benefit to your users. While this is not terribly significant to the overall speed of your network, DHCP does generate a great deal of extra network noise.

Thirdly, a static addressing scheme removes a server from the Mix. Without dynamic addressing, you do not need to worry about the DHCP server failing. The absence of a DHCP server eliminates one point of potential trouble in your system.

A Sample Addressing Scheme Using TCP/IP

My addressing scheme is based on the standard "Class C" network. This means that I can connect up to 253 devices to my network, with numbers ranging from 10.0.0.1 to 10.0.0.254.

Incidentally, the "Class C" network is the smallest of three classes of networks. The other two classes are referred to as "A", and "B". Class A and B networks are both strictly speaking, usable for a small network, but are better scaled to larger networks. They are not discussed here, because it would be typically inappropriate to use either a Class A or B network with a network of the size covered in this book.

Our network looks something like the following:

10.0.0.1	Gateway to Internet (our firewall with no DHCP server)
10.0.0.2–10.0.0.30	Our networked Servers in the server room
10.0.0.31–10.0.0.40	Our networked Printers
10.0.0.41–10.0.0.100	Our main building, computer lab computers

10.0.0.101–10.0.0.120	Our main building, first floor client computers
10.0.0.121–10.0.0.160	Our main building, ground floor client computers
10.0.0.161–10.0.0.180	Our school building, main floor client computers
10.0.0.181–10.0.0.200	Our school building, second floor client computers
10.0.0.201–10.0.0.220	Our school building, third floor client computers
10.0.0.221–10.0.0.230	Our Residential house client computers
10.0.0.231–10.0.0.254	Expansion space for future use

I maintain a database with the above map of numbers as the first section. In the subsequent sections, there are tables with a listing broken down, of each floor and the machines on that floor.

The database maintains information regarding the IP Number, the network name, and the current location of the system.

In the third, final section, there are templates for adhesive labels each with a unique IP number printed on each number. When a new computer is installed, the appropriate tag for that computer is printed, and placed on the back of the PC. The chief advantage of this scheme is that there is always a record of the remaining available IP numbers. Additionally, the numbers on the back of the PCs on the labels provide a record should there be a problem with the PC or the network.

INTERNET CONNECTIVITY

There are many options for Internet connectivity. All of these options presuppose that the network is using a TCP/IP addressing scheme. Using TCP/IP as the only network protocol makes Internet connectivity, especially high-speed Internet connectivity, much easier to integrate into your system.

CONNECTION TYPE	ADVANTAGE	DISADVANTAGE
MODEM/POOLED MODEM (LOW SPEED CONNECTION)	* Relatively Inexpensive * Easy to get service most anywhere, although dial-in number may be a long-distance phone call (Older wiring can also pose speed problems)	* Comparatively Very Slow * Not a good option if you wish to host services * Pooled modem increases speed, but increases cost and is not particularly reliable. * Access delay exists while waiting for computer to dial number
ISDN (Integrated Services Digital Network) (MEDIUM SPEED)	* Easy to get from local telephone company	* Very Expensive * Relatively Slow
SATELLITE DISH (MEDIUM SPEED)	* Provided that you can get service, easy to implement	* Hardware is not as reliable or standard as other options. Connections, especially, are non-standard for networking hardware * Requires a host PC to act as a gateway for services * Odd hardware arrangement precludes use of a real hardware firewall device * Support level from one company has been found by author to be inconsistant, and unreliable, especially on weekends
CABLE MODEM (HIGH SPEED)	* Very Fast * Easy to set up and support * Can be used with firewalls, routers, etc.	* May not be able to host services, per agrements with cable provider * Not available in all areas * Shared nature of network makes speed unpredictable, and subject to number of users in your area
DSL (HIGH SPEED)	* Fairly Reliable	* Typically not as fast for similar cost as Cable * Companies are less reliable than the actual service. Lots of turnover, no consistancy
T-1 (HIGH SPEED)	* Proven Technology * Very Reliable * Fast * Great for hosting Services	* Relatively expensive * Requires a great deal of involvement with telephone company * Complex to set up, requires networking knowledge * Interface hardware is expensive

FIGURE 2.19: INTERNET ACCESS CONNECTION ATTRIBUTES

SERVER ROOMS AND SPACE PLANNING

After you've decided on the network equipment, you need to plan the space and environmental aspects of storing the head-end equipment that manages your network. Equipment outages, and network problems tend to manifest themselves because of inadequacies in space planning.

Storage is not an adequate word, because this equipment is running continuously. When running a system, one would never opt to turn off the servers at night. Additionally, servers should be protected, and allowed to run with a minimum of tampering or adjustment. Insufficient space to properly hold the equipment in a reasonable way causes outages and service fluctuation.

Too often, spaces considerations for system rooms are done as an afterthought. The network servers should sit on fixed shelving or, preferably, racks that are attached to the floor. There should be open space as to allow access to the front, rear and sides of a server, and still have room to move about. Servicing equipment with insufficient room can cause problems, as you or the technician will be distracted or uncomfortable when working on the equipment.

Additionally, make certain that certain types of equipment have clearance. Ill-shielded video display monitors, especially, can reap havoc with network hardware. The equipment should be placed far enough apart so that air can flow through the chassis of the machinery, in the manner in which it was designed.

Standard 19"x72" racks simplify this process. You will likely have a rack or two around for the patch panels. Most installers require these racks to be fixed in place, so as to protect the installed wire from damage related to movement or stress. You can purchase additional racks, called "relay" racks, to which network hubs, routers, and other equipment can be attached. Shelving which mounts into the relay rack is available for equipment which cannot be mounted to the rack, such as display monitors, PCs, and other equipment.

Deeper racks are available for use with rack-mountable servers. The cases for servers that are specifically rack mountable are quite deep, much deeper than most ordinary PC server cases. Damage to the server case, or to the rack can occur if the server case is used with the relay rack. Servers should not be mounted into relay racks.

CLIMATE CONTROL

Poor ventilation or cooling decreases the life span of equipment. I have seen in many system rooms little or no ventilation. If a choice has to be made, air flow is probably a more key factor than cooling. Most equipment is equipped with fans, but certain network hardware, and video/audio hardware generates a great deal of heat which dissipates to the other equipment.

Possibly more immediately problematic in an installation is an inadequately implemented electrical service or lack of surge protection. Usually, this is not the problem of the electrician or designer, but of the end users and network administrators. Especially in situations where system rooms have been converted from other spaces, you will find that there are lots of "daisy-chained" surge suppressor outlet strips, all connected to a few outlets around the room. If this is the case on your network, it is a time bomb.

Power supply failure is the silent killer of computer equipment. Erratic problems manifest because the power supplies are not getting either sufficient or consistent levels of voltage from the power source. This can occur because of means outside of your control, which have to do with the electrical service in your area.

Planning a system room is different from planning other spaces in your physical plant. All too often key issues for system rooms are underestimated or forgotten, leaving expensive gaps to fill.

Certain factors that may guide your decision include the electrical service, the physical location, cooling and ventilation, and physical security.

Consider placing the system room close to an outside door on the ground level, so that it is easily accessible by service personnel. Bear in mind that it is by nature, a loud room. There is a great deal of white noise from fans, which will escape into the surrounding area. Locating the room into an office area where people tend to work is not generally an ideal location. If this is impossible to avoid, consider using acoustic tile in the system room to insulate the outside offices from the white noise of the system room.

When planning the room, if possible, install larger than standard-sized doors. Double-doors may be better, with a great deal of height, for accessibility. Also, try to locate the room near the electrical service points in your building. Certainly, the room needs to be in the same location as the patch panels, which by extension may be where the electrical service is located. This will make connecting all of your equipment much easier in the future.

As with many issues, your organization needs to specify your requirements. Never rely on the general contractors who are doing the building, or the architects who are doing the designing to read your mind. Unless the people who design your space are specifically educated in network and computer issues, there is likely to be some problems having to do with adequate cooling, lighting, electrical protection and service, and space.

In our system room, electrical service panels were installed directly in the room, so all of the circuits are available at close hand. There is a display indicator panel installed onto the circuit breaker box that shows if there is ground fault protection, and that the protection device is working correctly. In our installation, there are higher-amperage electrical outlets for use with battery backup equipment. The importance of battery backup equipment (also called UPS–uninterrupted power source–equipment), varies with your installation. We have battery backup on our two key servers and switches, so that the base network will not fail in the event of a power failure.

Although not a topic directly related to this book, our installation has a digital telephone system, a PBX. The telephone system relies entirely on electrical power to operate. In the event of a power failure in the building where the phone system is housed, the entire telephone system will fail at our organization, which is over 100 telephones. In this day and age, telephony is as important as any other life-safety equipment, we decided to install battery backup units that will power the telephone system for up to 3 hours after a power failure.

While in most areas, electrical codes are fairly straightforward, HVAC and space requirements are usually left to the architects and space planners. It is a good idea to talk to the HVAC contractor prior to any decisions about cooling a system room. System rooms should remain, at full operating capacity at between 68 and 73 degrees Fahrenheit. In certain situations, these systems may need to be sized with an idea of the amount and types of equipment housed in the system room.

Because our system room is in the inside back corner of the basement, there were no options other than to use a ductless-type air conditioning system. Many manufacturers make a variety of ductless systems, which are popular options for many system rooms. Mitsubishi manufactures a popular model for use in system rooms, called "Mr. Slim™". Sanyo and a variety of other HVAC and electronics firms manufacturer similar, small units which are ideal for such rooms. Ductless cooling systems can be installed a relatively long distance, usually to a room or space which has no direct outside access. Window air-conditioning systems and most traditional split systems like the heat-pumps found in most homes need to have an outdoor condenser unit within close range of the indoor part (the fan). Ductless systems, in general, are engineered with limited space in mind, so they are shaped to fit into the walls, drop ceilings and other cavities of a room. We have found that it is more expensive and time consuming to maintain our ductless system than with a traditional split system. This may factor into your decision of where to locate your system room.

As important to the life of your equipment as the cooling is to have adequate ventilation in your system room. Our system room is a long L-shaped room, we decided to locate the cooling equipment on one wall. Unfortunately, the cable installers located the wire patch panels on the other side of the room. We found that, after all of the servers had been installed, that "heat pockets" were developing, significantly so in the shafts created between the rows of racks. This was causing heat-related damage to the servers, so we installed several cooling fans using a metal grid, which was then installed, above the level of the racks. While this creates a significant improvement of airflow for the equipment, it is an added a great deal of noise to the room.

WIRE MANAGEMENT CONSIDERATIONS

Wire management is important in any organization, because a great deal of troubleshooting time is wasted because of inadequate labeling of wires, sloppy installation or wire management. Convoluted wire installations can ruin an otherwise well run network installation.

Historically, a popular option was the raised floor. Larger system rooms had raised floors made out of square carpeted-or hard tiles. These tiles could be lifted out, revealing a crawlspace under the floor. Wire management has changed, and today, smaller installations usually do not call for this type of wiring management systems. Still, if issues of either aesthetics or organization are important to you, you may consider installing a raised floor in your system room.

Network Server rooms are typically placed in a physically centralized location. They should have some form of conduit going out to the remote locations, both inside and outside of the building, if necessary. Main Server rooms are called, for the purposes of wire management, MDFs, or Main Data Frames. In this example, the server room is below grade, this means that is it is underground. The conduits are the tubes located at the top of the room near the ceiling. The conduits carry cable TV, telecommunications and network wire to jacks in the main buildings, or to patch panels located in other buildings, referred to as IDFs (Intermediary Data Frames).

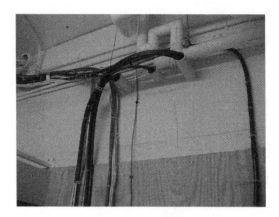

FIGURE 2.20: A CONDUIT ARRANGEMENT.

When the wire enters the room, it should be firmly attached to the wall, and to the rack. Wire is a permanent fixture, and the rack should be permanently attached to the floor, or to the wall, depending on the type of rack. A moveable or unstable rack can destroy the delicate interconnections between the wire and the patch panels.

The wire is ultimately connected, or "Punched Down" as it is referred, to connections on the rear of the patch panel. This is shown below:

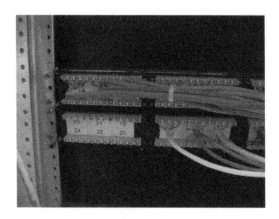

FIGURE 2.21: REAR OF A PATCH PANEL

The connections are then made from the front of the patch panel to active equipment. Active equipment means network equipment like Hubs, or Switches. This completes the connection from the jack on the wall in the office to the hub.

FIGURE 2.22: CLOSE-UP OF FRONT PATCH PANEL
The front of the patch panel is physically connected to the back.
Small cables, called Patch Cables, connect the jacks from the back of
The panel to the hub. This is done to simplify moves, adds, and
Changes to the system and eliminate any wear and tear on the
important Premise cabling.

PHYSICAL SECURITY ISSUES

Most important to the placement of a system room is the physical security of the location. Server rooms are prime targets for vandalism and theft, especially in areas that may be prone to crime. Often, a concentration of tens of thousands or hundreds of thousands of dollars worth of equipment is located in system rooms.

Theft attack is twofold. There are 'smash-and-grab' type thefts. The hardware is the target, and people steal only the software programs or hardware (computers, etc.) which appear to be valuable. This type of theft is obvious when it occurs, because things are missing, entry has been forced, et cetera. Fortunately, this type of theft is decreasing, as the value of this type of equipment drops on the secondary market. Alarm systems, the recording of serial numbers of equipment, off-site backup strategies, and of course vigilance on your part and the part of your security personnel helps to prevent the effects of theft.

Implementation of practices such as keeping the system room locked can help to reduce theft. In our organization, we do not even allow contracted cleaning personnel into the system room. While disallowing cleaning staff may seem like overkill, a seemingly innocent drill, floor buffer, or vacuum cleaner plugged into the wrong outlet can reap havoc on an entire system.

In our system room, we have tried to combat this issue by limiting the number of keys to the system room to the plant management team, the associate system person for the school, and myself. This limits the number of people who can access this room, and thus the amount of opportunities for theft. It also has the side benefit of limiting the number of hands in the mix.

The more serious and insidious type of theft is that of information. Our organization maintains lots of types of information. Information is kept on our internal operations and our bank account balances. Additionally, the school's grade system and children's health information are kept on the system. Our computer system maintains parishioner data, such as demographic information.

As you can see, there are a host of opportunities for the dishonest individual to access privileged information on our system. The machine left logged into the network after hours, the hastily discarded backup-tape, or the user account with no password all contribute to this problem. As you move towards mechanized data management, your organization will have these same issues. Prudence and a keeping a keen eye over your operation can help to limit the risk of these issues.

Theft of information can be thwarted by several simple practices. It is your users' responsibility to follow these practices, but you can set rules for the network that help to structure your guidelines.

- Every User is required to have a password

- Every User must have a discrete user account, which lists their name

- All machines must be logged out and shut down prior to leaving for the day

- Machines should have a "BIOS" level password. BIOS passwords prevent idle people from being able to start up the computer at all.

- No passwords should be left on sticky notes or in other obvious places. Check for this periodically when you go to do service.

SPECIALIZED LAN EQUIPMENT

LAN (Network) Equipment is equipment that was specifically designed, or modified, to work over the network connection. There are many types of specialized LAN equipment, often this equipment is a re-engineered version of an earlier developed product. Examples of LAN equipment are networkable hard disk drives, printers, CD-ROM drives, or scanners. Using LAN-enabled equipment can add rich services for your users, and can make using and maintaining your network easier.

Network Attached Storage and Backup

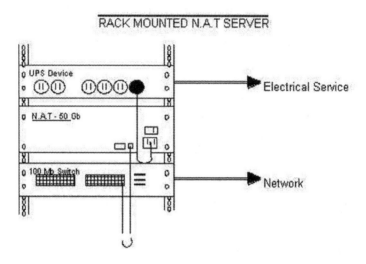

FIGURE 2.23: NETWORK ATTACHED STORAGE

Network Attached Storage devices are basically large-capacity hard disk drives, or arrays of hard disk drives, that have a network adapter embedded into them. These pieces of equipment are seen as a file server to other computers on the network, but are much cheaper to install and are easier to maintain than a real file server.

Network attached storage equipment is about the size of a home VCR, and connects to power and LAN connections. Although 80% of the hardware is the same as a complete server, there are differences. Most NAT systems are based in UNIX operating system environments of one type or another. Often, there are no ports, video cards, keyboards, or mice available for NAT equipment. This equipment is controlled remotely through a web-page interface, and thus, can be controlled from any PC on the same network as the NAT device.

Similarly, network backup devices are tape-backup drives which have network controllers embedded into them. They contain power and network connections, and have no local controls.

The limit of the control or monitoring that can be done locally on most of these devices is through one or two power/activity status indicator lights.

NAT and Network Backup is sold as a Plug-and-play option for smaller networks, or for expansion to larger installations. UNIX, the basis of most of these systems, is a very stable system, and is not prone to failure.

ADVANTAGES OF A COMPLETE SERVER
Easier to Fix onsite
Control/Configuration is greater than with NAT
Expandability is greater than with NAT

ADVANTAGES OF NAT
Less expensive than a full server
Less prone to failure
Easier to Implement

Networked CD-ROM servers

Networked CD-ROM Servers, sometimes called CD Towers or Jukeboxes, are shared CD-ROM drives. These systems are usually housed in a large server-type box, and include a hard disk drive, and a network adapter. Depending on manufacturer and model, these systems may be a complete server, with monitor and keyboard, which can be configured for use with your network operating systems.

A real CD-tower is a system with at least 6 CD-ROM drives, although you can build a simple CD-Tower with fewer, if you so choose. CD-Towers can have as many as 30 CD-ROM drives, and can be housed in large cabinets with multiple power supplies, fans, and other server-grade redundancy features. As with any other computer hardware, you will want to scale the system to your needs.

Shared Printers

Shared printers will be the one thing which you will probably decide to implement on your network. Every modern operating system allows for the sharing of printers via people's computers. Even older systems allow specialized "Print Servers" to be installed on computer CPUs, and thus allow printers to be shared.

Sharing printers means that much better quality printers can be distributed, and shared by groups of people. More expensive Laser Printers are in the long run, cheaper to maintain. They produce much higher quality documents, and are more functional and effective in a business environment. Inkjet printers should be avoided, as they say, like the plague. The prospect of removing inkjet printers from those who already have them may be disheartening to some people. These people may make an argument to install more, albeit cheaper ink-jet printers, rather than the laser printers. Disregard this argument, because it is a terribly shortsighted direction to take. Printers are highly mechanical and unreliable pieces of equipment. Rugged, well-made printers will save your organization from having to prematurely replace equipment in the years to come.

You need color, you say. Well, short of buying a color laser printer, you might try this option. Use a Printer Sharing device to attach a normal color inkjet printer to the network. Place the inkjet printer in a central location, so that it can be used by anyone on the network in that area.

Printer Sharing Devices are small boxes that are made by certain printer manufacturers for use with their printers. These boxes connect to the power line, the network, and through a parallel or USB connection, to the printer itself. They are usually configured via a Web Page interface, and can be controlled via any other networked PC. For the purpose of security, they can be password protected to prevent unauthorized tampering.

One thing that should be avoided is the tendency to set up printer sharing arrangements from computers used by individuals for purposes other than sharing the printer. Connecting an ink-jet printer to some-one's PC and using it as a pseudo-print server is a recipe for problems. This person will shut their machine down in the afternoon, the machine will crash, and other problems manifest themselves. Ink-jet and non-networkable laser printers should always be attached to spe-cialized network attachment devices.

UPS and Surge Protection Devices

Although not providing direct network services, the UPS (Uninter-ruptable Power Supply) or Surge Protection device are crucial parts of your network. Each server, laser printer, and client machine should be connected to at least a power surge protector. Power surge protectors are described as such on their packaging, and are more than the $5.99 power strips which are available at most office supply stores. They should be of a reputable name brand, and have a circuit breaker or fuse that can be reset or replaced. Power surge protectors that are of high quality will offer a registration, and a guarantee on your equipment. Most installations are not immune to power surges, and lightning. More detrimental to network equipment than a power surge, is a low power situation. Brownouts (low-power situations) are more common in some areas than others. The unprovoked dimming of lights is symp-tomatic of this type of problem.

Production Servers and network equipment should additionally be connected to a battery backup device, called a UPS. Battery backup devices connect between the electrical service outlet and the power connector of the electronic system. They provide a greater protection, because they prevent an unplanned power outage of a piece of equip-ment. This can prevent software and hardware failure for a variety of reasons. Most modern UPS equipment can also communicate with the server software, and can be programmed to shut down the computer system properly rather than abruptly, and without warning to the server. This feature is a great asset and can prevent data loss and damage to hard disk drives, backup tape drives, and other mechanical equipment within a system.

CLIENT PCS

Client PCs are different animals than Servers. Even the operating systems differ. In certain systems, such as many UNIX variants, there is little or no differentiation between a file server and a client, but in most all other major operating environments, differences exist.

In a typical PC, the client hardware should be more performance oriented than in an equivalent server. We said that servers should be input-output oriented, clients should be user oriented.

Being "user oriented" means having adequate video performance for the size of the screen, having adequate sound and speakers to undertake the applications for which they were designed. Being user oriented primarily means being responsive and usable, and being designed with the user, and the office environment in mind.

Client computers should be quiet, and should be purchased with office furnishing in mind. Computers should not overpower a person's desk or work environment. Large towers, desktop computers that eat up a quarter of the desk space do this. In the same way, loud computers can be disruptive to the work environment. In most networked environments, the need for equipment duplicity is reduced, and most computers can exist with a minimal amount of drive hardware. CD-Writers and DVD Writers, and other specialized hardware can be put in central machines designed for these purposes. When dealing with client computers, the rule of thumb should be to give the user what she or he needs, and no more.

This is a key problem with regard to most off-the-shelf computers. Especially when dealing with consumer-oriented offerings, there is the unfortunate tendency to add lots of add-on hardware. Off-the-shelf computers may come with pre-loaded diagnostic and productivity software. If you have standardized, or purchased site use licenses for a particular program, the pre-loaded software may be of no use to your organization. My advice is that if you insist on buying a computer this way, clean everything off, and start from scratch.

If you are planning on using Windows, as many organizations do, there is a decent option available. Go to the store, and purchase an in-the-box copy of Windows. Your computer will likely arrive pre-loaded with Windows, so you should have a legal license for that machine. You can then install a clean version of windows on each computer, provided that the license that came with the computer matches or is newer than the version that you are trying to install.

If your computer manufacturer indicates that this arrangement will not work, or will void your service contract, move on to the next vendor. There are certain vendors whose computers will not work correctly with off-the-shelf versions of operating system software, because they require certain specialized software to work correctly.

This may be satisfactory for your servers, but it is never necessary for your client PCs. When dealing with client computers, You want to be able as little as possible. When doing a product install, give the minimum required to be functional. There is always the ability to install more, if necessary. Installing programs or extensions which have no use will make the system more complicated to use, and more importantly, to troubleshoot. The rule should be: Keep It Simple.

Standardizing the Selection of Programs and "Look and Feel"

In the same way, your client PCs should be standardized. If you are running a version of the operating system or program, to whatever degree is possible everyone should be running that same version. As with other aspects of the client PC, there should be no perceptible difference between one computer and the next. Without biasing this book, these wars have already been fought, and at least for the next few years, your decision about operating systems and productivity software has likely already been made.

Although there are better alternatives, it is important to maintain an even playing field with what your users likely have at home, and what your neighbors have.

Standardizing your software as much as possible makes training and troubleshooting much easier, and will make the environment familiar for your users, should they need to switch PCs, or work in another location temporarily.

The look and feel is the color scheme and the desktop background color or picture. While I do not advocate locking these choices down, I do think that there should be default selections in every network system, so that you can make accurate screen shots for handouts and training documentation for the users.

Standard Versus Proprietary Architecture

When purchasing PC equipment, there are really two types of choices. You can choose to go with a generic system, or you can go with a name-brand manufacturer. Name Brand manufactures, are recognized companies such as Apple, Gateway, Hewlett-Packard, and IBM. These and other similar companies are at the forefront of technology, and engineer, rather than just assemble computers. These firms address through design how they can make a computer work better and more efficiently. They will likely fabricate some parts and modify others to meet their specifications.

The Result? With a commercial grade of system you will get a better computer from a name-brand manufacturer, every time. The computers which are developed by companies like those listed will put many man-years of engineering into making sure that their products work well with standard accessories and networks, and that the components in their systems work together.

The downside to this is that you will never be able to walk down the street to the local computer store and fix the broken computer. Problem resolution will always involve calls to the vendor, and possibly a shipping arrangement or a replacement PC. If you are like me, I would rather tinker with a computer than wait on hold, and as such, I buy very few name-brand PCs.

With a generic system, life is a little less consistent. Computer systems from local vendors are not engineered in the same sense. They are assembled from prefabricated off the shelf components. Let's say your name is Joe. Your name is "PC Joe", of PC Joe's Computer Store. You build and sell computers and accessories out of your store for a living. One day, John Q. Customer walks in, and wants to buy some PCs. When you get a contract to build, like 10 or 20 computers, you look on the Internet to get a good price on mainboards, sound cards, hard drives, cases, screws, all the parts which go into making a computer. Unless Johnny Customer tells you that he wants ABC brand Network Adapter cards, you buy the cheapest network adapters which you can find, because markup is where its at, baby.

Here's the problem. You don't know ABC network cards from ABC gum. You may have some personal experience with ABC. Maybe you buy their sound cards or hard drives. You may have a friend whose friend is a network admin, who tells you to "stay away" from a certain brand. Most likely, your level of knowledge and by extension the level of "engineering" of your systems is based on your own experience and on trade advertisements and articles about performance. You have no choice but to rely on the fact that the Taiwanese company that makes the network adapter is complying to the standards which dictate network cards. If this is the case, you're OK. If not, you'll have 10 or 20 computers which crash a lot, and Johnny will have some more gray hairs on his head.

Realize that, in most situations, generic is an acceptable way to go. The parts *are* standardized and play by the rules, and while most reputable dealers do not know brand ABC from XXX, they know what to look for in standards, and they deal with enough equipment to know what to stay away from. The key is to find a dealer, preferably one who's been around for a few years, and to initiate a relationship with this person. Buy a few PCs from a company you've heard something good about, and let them put them together for you. See how these machines work out. Do the power supplies crap out after three months of use, do they crash a lot more than your other PCs? To help keep things honest, you may want your users to keep a log describing their computer's reliability.

If you find that this company makes a good PC, and you may wish to do more business with this individual. Remember what I wrote about the vendor relationship previously in this book. Small, local vendors are usually a better bet when you have problems, or when you wish to set up accounts or the like.

Peripheral Equipment/Special Purpose PCs and Equipment

Important to realize is that your users will require extras on their machines. They may require that certain PCs have scanners, or better video cards or larger displays.

Typically, you will want to stick with a seventeen inch display, as this size is adequate for most purposes and computers, but is not so huge as to cause visual problems or to have issues with flicker, as some of the larger screens do.

Computers that are used in desktop publishing or web publishing activities may require larger screens, up to 20" or 21" in size. These computers will require premium accelerated video adapters, and additional hardware, they may also require scanners or color printers.

Special purpose computers may also include training PCs with projection equipment, or computers with digital video capabilities.

Envelope Printers

One recent development that is of great utility, especially to a small organization is the envelope printer. Coupled with a postage meter and folding machine, this ensemble can make it possible to produce the kind of professional results previously available only from a professional printing service.

The envelope printer will have some form of a hopper, which allows the blank envelopes to be stored. The printer is configured to print on standard sizes of envelopes, from data stored in a database.

Envelope printers can be connected to either a network, for shared printing ability, or directly to a PC via a parallel (Printer) port connection. Key to choosing an envelope printer is to make certain that it will work with your database, and that it is compatible with your operating systems. A printer that uses a generic print driver, such as a "text-only" driver is the most versatile option.

Photocopier-Printers

As office automation equipment becomes more integrated, one device that provides economy and coordination is the shared copier/fax/printer. This device allows some of the following features, contained in one floor-standing, or tabletop model. The key advantage is the reduction of equipment, and the decreased cost to support and print. Although copier-printers are more expensive than either a stand-alone laser printer or a stand-alone copier, the added functionality is more economical than a separate printer and copier. The cost to support, as well, may be reduced, as there is one service contract for the combined device rather than two separate contracts.

- Shared Laser Printing, either monochrome or color

- Scanning, either in grayscale or color

- Facsimile, shared facsimile services

- Document server. Shares documents for reference purposes

- General Copying capabilities, collating, etc.

DONATIONS

Whether dealing in software, hardware, or telephone systems, you will encounter a phenomenon in the non-profit arena: The Donation. Donations can be a panacea or a Pandora's box, depending on how the practice is managed. People long associate donations with non-profits, and schools, because it is a good way to get rid of aging equipment, and it is a means of getting tax credit. Yeah, there is some altruism in there, sometimes, too.

I have worked in environments where the organization expected to be able to run a system more or less completely on donations. This is very real, and doable, given some basic planning and a little knowledge of older hardware.

Avenues for Acquiring Donated Equipment

There are basically two types of direct donations of hardware. These are donations by individuals, and those by corporate donors. Corporate donations are inherently larger in scale, and are easier to deal with in some respects.

In a good economic climate, firms will upgrade equipment regularly, and a fleet of relatively new equipment needs to be removed to make way for the "state-of-the-art". A good relationship with local corporations or universities in your area can help to bolster your technology infrastructure. Companies and IT department managers are rarely inclined to keeping old equipment, as space is usually at a premium. Dumping computer equipment is not a viable option, as it is costly and environmentally unsound. If you maintain communication with the people who plan and schedule technology upgrades, they will begin to look to you as an answer to this unending problem, from their perspective.

Corporate donations are not always a good answer, and there are several pitfalls. The key problem with corporate systems lies in its advantage. Corporate IT equipment is usually comprised of a quantity of the same exact system, or similar systems, which have been ordered by the firm to meet their needs. *Their needs may not meet your needs.*

Systems may all be configured with network adapter cards, where you may need systems with modems. Unlike individual computer owners, who regard their computers as "major purchases", companies are less fastidious, and rarely save any documentation, boxes, or diskettes that originally came with the system. As such, trips to Internet websites will become a reality for information and driver software. Asking a lot of questions about systems and configurations, and doing research about what they wish to donate up front is a wise idea.

Depending on your funding level, running your system on donated equipment may be your only option, but it requires a lot of discipline, communication, and the willingness to throw things away which are not part of your overall vision or system.

Whether dealing with companies or individuals, creativity, patience and knowledge of many older systems is also a requirement. Often donated equipment will have no manuals or drivers. Depending on the donating party, a trip to the computer store for parts may also be necessary.

On the whole, donations from individuals are a different game from those of corporations. More often than not, you will find that there is a compelling reason why the person is donating, rather than selling or trading in, their old equipment. Many individuals would rather donate than discard old computer equipment. Computers owned by individuals are usually not as "commercial grade" as those donated by companies. While individuals usually donate the boxes, printers, and literature that came with the computer originally, these machines are usually designed not with a working environment in mind.

To filter what we receive, both by organizations and individuals, our organization has developed and published guidelines that are to be forwarded to anyone wishing to donate, either at the personal or corporate levels to our organization. These guidelines discuss thing such as baseline, or minimum, specifications for donations. They also list specific equipment that we are currently looking for, such as data projectors and network infrastructure equipment.

Use of these types of guidelines may seem arrogant, but it is necessary to be up front with people. There is little value in taking a computer just to dispose of it.

This being said, older computers still have parts, many of which can be used to repair machines of similar vintage. Just try finding a 500 Megabyte hard disk drive, or an ISA network adapter card with a twin-axial connector. Finding "vintage" parts is becoming more difficult, and more costly every day. Some older equipment must be purchased through alternative means, like online or through surplus parts suppliers, which can be unreliable and expensive.

Expectations and Reality

Once you have made the decision to make donated equipment a part of your system, you should do some planning.

The reason why donations are so widely available has to do with two things. In a good economy, there is a push by companies to put a lot of resources behind information technology. In a bad economy, the amount of available resources will decrease, and donated equipment may become available less frequently, resulting in older equipment, on the whole.

Realizing that your system will be chronologically behind, you can weigh the advantages and disadvantages.

In the era of the writing of this book, there is a window of about a decade, spanning back until about 1990, when the equipment will be reasonably compatible, and reasonably usable.

I say this, because Windows 3.1 came out in 1991. I have maintained systems comprised only of Windows 3.1 systems, and software. Many organizations continue to run software of this era, with a great deal of success. With a clever system administrator, or volunteer effort, computers and printers can be set up successfully, with Windows 3.1

software. Even internet connectivity is an option, and most major software, if it can be found, is available for this platform.

The key to maintaining a system of older equipment is to have a consistent era of equipment. If your aim is to have a system of Windows 3.1, or Windows 95 based computers, make sure that all of the computers are of this era. As with a brand-new network, This will make troubleshooting and training easier for you.

Bear in mind, though, that while your older systems will be able to handle the software of their era, newer software will not, in all likelihood, work with this arrangement. You will need to find programs, such as word processors, spreadsheets, et cetera, of the same era as your equipment. Software, and operating systems can readily be found online, and on online auction sites, such as "Ebay". In addition, swap meets and computer shows are loaded with older versions of software

In the world of the donated PC, here are a few types of system which you are likely to meet: The Windows 95, 98 or NT 4 machine, the Windows 3.1 machine, the DOS machine, and the Macintosh. There are some others, such as IBM OS/2, and UNIX, which you may run into as well.

In all likelihood, considering a corporate donation, a company may be retiring an "era" of systems, and a large quantity of similar machines may be available. This may aid you, as software and hardware support equipment may be available as well for donations.

Below are a few types of machines that you may meet in the course of a donation experience.

The PC Platform

Windows 95, 98, and Windows NT

Newer versions of windows are relatively easy to support. Hardware is not that old, and most systems can be readily repaired. Swap meets may be an option for finding older software, or online. To most computer literate people, these systems are familiar. One word of caution: Computers and equipment configured with Windows NT operating systems are notoriously incompatible with other networks, even those of the same type. If you receive computers with these systems pre-loaded on them, it may be necessary to remove the Windows NT system, and to start from scratch.

Windows 95, and 98 are very versatile. They will run most all new software, and will accommodate add-ons like video capture cards, CD-Writers, and CD/DVD drives. The list of available software is not provided, because software is available off-the-shelf which will run on these systems. It would behoove you, though, to stay with versions of productivity software which were new when the systems were new. After windows 95, many software packages started versioning themselves using the year. Look to this as a loose key regarding the age of a program. Computers such as Intel 486-based machines will run Windows 95, but may not run new versions of Microsoft Office, or other major commercial packages.

Usually, an Intel 486 or Pentium based computer is required to run Windows 95 acceptably, with at least 32 Megabytes of RAM. Windows 98 requires a faster machine, of at least a Pentium 233, with at least 64 Megabytes of RAM. Using Windows 95 or 98 as a base for your systems is the best option for most small networks, as it works well with modern networking standards, and eases hardware installation.

Windows NT 4

The system requirements for Windows NT 4 are similar to those for Windows 95. The appearance, and software facility is also similar to Windows 95 or 98. NT is a very stable system, but while not more complicated, is different in its manners from DOS-Based Windows systems, like 95. The hardware management and network management differs significantly. The only downside to an NT system is that it does require some more maintenance and set up, and it is likely that machines retired from some other network will not work correctly at your site, without having the operating system reinstalled, from scratch.

Windows 3.1:

Windows 3.1 saw an explosion of software for windows. For over five years, all major companies were rushing to put out graphical software. Look for the "Windows 3.1 Compatible" logo on boxes of software. Microsoft Windows 3.1 is best run with Microsoft DOS versions 5.0, or 6.0.

Windows 3.1 requires an Intel-based computer with at least a 386 type processor, of at least 20 Megahertz, with at least 4 Megabytes of Ram. A hard disk and a 3 1/2 inch high-density floppy drive is required to install the system.

Some Major Applications:

WordPerfect 5.1 for Windows (Word Processor)

WordPerfect 6.0 for Windows (Word Processor)

Lotus Ami Pro for Windows (Word Processor)

Lotus 1-2-3 for Windows (Spreadsheet)

Harvard Graphics (Business Graphics)

Aldus PageMaker 5.0 (Graphics)

CorelDraw (Windows 3.1 version) (Graphics)

Microsoft Office 4.3 for Windows (Productivity Package)

    Word version 6.0 (Word Processor)

    Excel version 5.5 (Spreadsheet)

Powerpoint versions 4.0, 3.0 (Business Graphics)

Access version 1.0, 2.0 (Database)

DOS

The DOS system has ruled the computer world for over two decades, and it is a viable, stable platform for computers which serve end-users. It is not recommended as a server platform, as very little easy-to-implement DOS-based server software is available, even for use with small networks. Even new computers can run DOS alone effectively, although most newer hardware, like CD Writers and the like, have no programs which work with DOS. Some networks, also, may have problems coexisting with DOS-Only systems. The key problem with using DOS now was always the problem with it: The programs are all fairly rudimentary, but they are not consistent, and no knowledge can be ported from one to the next. Internet connectivity capability exists, but without a graphical environment, like Windows, it is limited to text-only access of websites and email.

IBM, and a host of other companies have been making DOS PCs for over twenty years. With some leg-work and creativity, viable systems can be set up with these platforms, which were basically constructed with business in mind. Even the oldest group of PCs can be networked together, given the correct hardware and software. DOS, however, is not a particularly good platform for Internet access. DOS is losing popularity, as more and more people are not versed in its operation. Like the Apple II, you may find people who know how to use DOS, but as most people joined the computer revolution after Windows became popular, more often than not they are not the normative.

A DOS only system would likely make little sense in most environments today, as the hardware would have to be ancient in order to not run Windows. On the plus side, DOS is pliable, and will run on any "PC"-type computer.

<u>Some Major Applications:</u>

WordPerfect 5.1 for DOS (Word Processor)

Lotus 1-2-3, many versions (Spreadsheet)

Harvard Graphics, Many DOS versions (Business Graphics)

Dbase (Database)

OS/2, UNIX, and Others

You may find, in your experience with donations, the odd UNIX or OS/2 machine. Some businesses and universities went with these operating systems because of their unparalleled stability and security. They both were, and are still, good systems. They are also both relatively obscure in today's market. Support for OS/2 is limited, and the applications base is much smaller than with the Mac or Windows. UNIX, while having a much larger base of software than either Windows, OS/2 or the Mac, is very difficult to administer. If you have the resources or knowledge to support it, a UNIX variant can be a great way to go, because it is versatile, stable, and secure, and it works well with networks. Most importantly, unlike any other major platform, many UNIX variants and applications software are free.

The nature of UNIX is that it be made available to anyone and everyone. People take the software, and the source programming for pieces of the software (called "Source Code"), and are allowed, even encouraged, to revise it, improve it, and redistribute it, as long as no charge is made for the derivative product.

You're probably not a software developer, but what does this mean to you? Unlike Windows or the Mac, where each computer must have a licensed copy of the operating system, which must be purchased, or donated, some versions of UNIX can be had free of charge. You can get them with a copy of a book regarding their use, from a bookstore or online. UNIX is difficult to administer, but no more so than any other operating system. Depending on the version, it can be made to run on any age or brand of computer, and works inherently well with networks and the Internet.

Apple Platform

The Apple II series

The Apple II series was, and remains in some circles, a staple platform. Apple II's are becoming less common donations as time progresses. Unless you have a compelling reason, like a specialized training package, you do not want to become involved with the Apple II series. Apple II's used to be a very viable platform, and tens of thousands of quality software packages exist for this platform. Unfortunately, very few IT people in this day and age really understand the Apple II series, and parts are no longer readily available. Additionally, these machines do not have the power or expandability to perform tasks considered commonplace on other, newer systems, such as Internet access or email. This is also not generally a good platform because few people remain who will be familiar or versed in its operation.

The Macintosh

The Mac has been around for almost twenty years, and is an excellent platform. A complete Macintosh of virtually any generation can be a viable computer to do a variety of tasks, and with a little research on the Internet, even Macs that are a decade or fifteen years old can be made to connect to the internet. The graphics capabilities are sufficient for many varied tasks. The key to success with the Macintosh is to building a complete system, as especially the older hardware is not compatible with anything in the PC/Windows environment, such as printers or software. The Mac has had several major transitions in the last several years. New off-the-shelf hardware for the Mac platform sold today will not work in older, Macintosh systems. As the Mac changed rapidly, and significantly over time, A working knowledge of not only the model, but also the generation of Macintosh that you are dealing with is crucial to success with this platform. Typically, Macintosh systems show up as donations in organizations who choose to move or remain with that platform.

One caveat: Integrating a Mac, especially an older one, into a PC organization, while possible, is generally a bad move because compatibility and transfer of information will forever pose problems.

Some Major Applications:

WordPerfect for Macintosh (Many Versions available) (Word Processor)

Lotus 1-2-3 for Macintosh, many versions (Spreadsheet)

MacDraw, MacPaint (Claris, now defunct) (Graphics)

Adobe PageMaker (Page Layout)

Filemaker (Database)

Users' Concerns

There are different types of users, too, who may require different types of equipment. Some of your users may require newer equipment, simply because their activities are more intensive. Donated equipment will begin to show its age with people who need more advanced resources, like the presence of a USB port, or a lot of RAM, or a newer processor. In most cases, it would be inappropriate to give these users older equipment if newer equipment is available.

In too many organizations, equipment follows the structure of management, as if a new computer was like a new desk. New equipment goes to the head of the organization, whose old PC is given to the secretary. The secretary, in turn, passes his or her machine down the line to someone else.

Unfortunately, in many organizations, the secretary and accounting departments use their machines to a much greater degree than those in management. The presence of a newer computer technology on the secretaries desks may make their jobs easier.

CASUAL USERS

Casual users are people whose jobs do not directly interact with the technology. Your plant staff, or associate people may fall into this category. Putting the proverbial "supercomputer" on these people's desks may be a waste of resources. They should be able to use the integral systems, like email and group calendar management systems. Word processing, database, and spreadsheet functionality may also be desirable.

INTERMEDIATE USERS

Intermediate users are those who interact with the technology in some ways. These may be people who type letters or who use email extensively to communicate. These are your computer literate users, and are the bread-and-butter level people in many organizations. They should have reasonably new equipment, but will probably not utilize too much specialized equipment if it is present.

ADVANCED USERS

Advanced users are the secretaries, and those who really depend on the operation of the system to do their jobs. These are your leaders in the technology sense. They should always be at the forefront of the technology upgrades, and should have access to the fast, new machines, good keyboards, and large displays.

Managing Donations

The only bane of donations is becoming a convenient "trash can" for other people's junk. The word 'Junk' may be derisive, but if the equipment is not usable to you, it is likely that you will funnel it to the trash—most organizations do not have the capacity to re-donate to others.

There are two ways to avoid this problem. Equipment should never be dropped off at your location, without prior plans or approval. Equipment which is broken should not usually be accepted, unless you have a system in place to do repairs, or unless it is a special circumstance. Your organization will need to evaluate what may be work repairing. Some surprisingly good specialized equipment, like color laser printers, projectors, good scanners, and other equipment can be had for free, if your organization is willing to have it serviced.

Equipment should also be evaluated on the basis of whether it meets your organizations needs. If your organization is looking for tower computers (the type which sit on the floor), and a company offers you only those which sit on the desk, you may wish to concede, and get the donation.

However, if your organization is looking to set up a Windows NT network, or a Windows 95 or 98 system, and is being offered twenty-year old IBM equipment, it is likely that you will not be able to realize your goal. This is likely a waste of their time, as a great deal of effort is required to donate equipment. It is a waste of your time as well, as you will have to give it away or dispose of it.

Typically, there might be a some factor which drive the decision of what type of hardware to standardize on. Perhaps your volunteers' or your competency is with Windows, or with the Mac. Perhaps the program you are using, or want to use, is not compatible with new system software. After deciding on the platform, you would want to do a few things so as to structure donations to meet your goals.

- Define written guidelines for donations, based on platform, processor speed, RAM, hard disk space, monitor size, and brand

- Build a knowledge base of older system software and support options

- Build a repository of resources and information for older hardware

- Check online for items such as software drivers, older manuals, and the like

Managing Donated Equipment

Unfortunately, there is sometimes a tendency to become sloppy with donated equipment. Especially in organizations where donations do not make up a significant part of the system, there exists a potential for lost opportunity and mismanagement.

A good first step to prevent this is to create a database of equipment which has been donated. This may begin, in part, because a letter is usually expected by the organization or individual who donated the equipment. This may be for tax purposes, or to reconcile their own inventory. Organizational and individual donation protocol does differ, but there are some similarities. Any equipment that your organization receives should be entered into a database noting several criteria. For each piece of donated equipment, this would require a new record in the database. The database should not be integrated with your main inventory, as some donated equipment may not work or be implemented into your system.

- Contact information for the party donating the equipment

- The Type of Equipment

- Brand Name

- Serial Number

- Date of donation

- Approximated Value

After a donation, you would want to send a letter to the organization or individual, thanking them for the donation. This letter should be on letterhead from your organization.

I send out two letters, one to thank the individual, and one at the end of the calendar year with an inventory of everything donated by that individual or organization. On the latter, I include all information from the database, including serial numbers where available. If the individual requests, I will include estimated values for tax deduction in the letter.

IMPLEMENTATION

At this point in your journey, you have moved well beyond your planning to the active phases of implementation. You may be in the process of establishing positive relationships with your vendors and contractors. Because hindsight is usually better than foresight, you may be changing your mind on certain things, revising and tweaking the models you originally set forth.

YOUR ROLE-OUT PLAN

Roll-out plans should be developed and codified before you begin any move to installation. The Roll-out plan is basically an expanded schedule, which describes days and/or time frames for upgrades and installation.

Communication is one area of technology management that is often overlooked. The roll-out plan is somewhat for your benefit, to keep you on schedule. In addition, it makes other staff aware of what is happening, so that they can plan accordingly. Your roll-out plan will probably include two types of documents. The first is your internal schedule. Secondly, a set of documents that will be sent out for communicative purposes. These memos should be straightforward and concise. They should also specify that unforeseen complications do occur sometimes, and that these circumstances sometimes have an affect on the schedule.

Typically, the first stage of a roll-out plan is to develop a simple schedule. Depending on the nature of the project, you may need to contact your vendors to schedule their parts of the project.

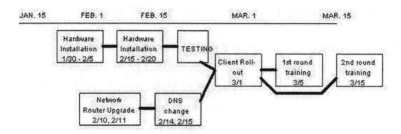

FIGURE 2.24: SAMPLE IMPLEMENTATION SCHEDULE WITH DEPEN-
DENCIES

The above diagram shows a basic roll-out plan. The first tract (top 3 blocks) have to do with server installation. Parallel to this is another tract (the 2 blocks below), which do not relate to the server install. The dark lines show dependencies in this situation. For instance, the client roll-out cannot occur until all of the previous activities have been completed. In this situation, though, the second round of training can indeed occur without first completing the first round of training.

In tandem with the organizational chart, periodic memos should be distributed to all of the people in a department of group of offices whom are being scheduled for an upgrade. This is especially important if the upgrade will mean to them a periodic service outage. Planned service outages should always be *proactively* disclosed to anyone who is going to be affected. This allows users, in turn, to plan for this down time. If a change in your plan occurs which is pertinent to your users, communication regarding this change should be distributed to the others in the organization.

As a network administrator, you know how reliant you are on the technology. Others in your organization also rely on the availability of the services you provide. Unexpected outages can work to undermine the efforts and (perceived) advantages of the upgrade.

Post-Installation Care

After the completion of any rollout, training should be promptly scheduled for your users. Part of your rollout plan should involve you getting familiar with the mechanics of the upgrade or installation to the degree necessary to answer questions and give basic levels of training to the rest of the users. In addition, users should receive documentation about the basic feature usage for the product. The documentation should be concise, clear, and should avoid any extraneous information about the project. This is fundamental training to begin using the product or service.

Later on, depending on the importance and scope of the project, you may wish to involve a third party educator. This is a person or company who is versed in the software, and will be able to teach more advanced concepts. Training issues are described in more detail later on in this book.

USER ACCOUNT PRIMER

User accounts are keys to the system, just like you have keys to your car. Just as you wouldn't leave your car running all day and all night, waiting for you to get in and drive, It is not a good idea to encourage users to leave their computers logged in all day and night. Network user names allow the administrator to set limits on what people can use, and to guide them to the appropriate avenues based on what they should be doing. This may seem impersonal, and to some extent it is. One of the big barriers to going with a network is that it is perceived as an embankment of power.

Decisions like who can access what information ideally do not originate with the network administrator. Organizational management people, or the manager of the employees in question should work with you to determine appropriate levels of access. This makes it a lot easier for you to manage the network, and to keep the peace with your users.

User accounting can control and grant access to two types of things. Service level features, like Internet access, email, and printing ability. It can also be used to grant or deny access to data, using shared directory security. A good mix of physical security of your servers, and user-level security can guarantee that the data privacy is maintained on your system.

Key issues with user accounts are privacy of data, and control. People who are not savvy with policies of the company come to see the network people as somehow lording power over others, using the security features of the system as leverage. It is generally understood that the administrator can see everything that exists on the network.

This does not mean that the administrator has no ethical responsibility not to look. Many newer network systems implement security, known as encryption, which can be configured in such a way that system administrators *cannot actually see users' informational files,* but can manage these files nonetheless. If you perceive that there may be a problem in your organization, you should investigate this as a possible corrective action.

USER ACCOUNTS

System administrators have a key role in the management of the users' accounts and their structure. When your network servers arrive, they will be configured with one or two accounts, the **Administrator** and the **Guest**.

The Administrative Account

The administrative account is called different things in different systems, but they all perform the same basic functions. Sometimes, there are administrative accounts for each of the servers, and a main administrative account for the network. Either way, the administrative account allows access to all of the files on a computer. This means that the administrator can install new programs or services on the computer, remove programs, and change other users' settings. The administrator can run and change the backup program on a server, and can move, delete, and read files belonging to other users.

This is a necessary function. File servers, especially shared drives that allow access by groups of people, tend to become electronic "junk drawers". We all have these drawers, usually in the kitchen, where everything and anything finds a home. There are half a dozen "mystery keys", many of which open nothing that you currently own. There is also a collection of junk mail from the last three months, and God knows what else. Some of the things in this drawer are actually important to you, like the extra set of keys to your car. Others could easily be trashed. The problem is, *who owns the stuff in the drawer?*

Well, when you look at everything individually, it can usually be deduced who owns it, but as a whole, it just looks like a mess to you. Maybe your daughter is saving that coupon for the music store, or maybe your husband actually wants to save those keys to the house you owned a decade ago. Nostalgia? Who knows?

As a system administrator, you will find that you have the same problem, but instead of three or four people living in a house, it's more like thirty or forty. Data management in these situations calls for one of two choices. Either you can define rules about use of the space, and police them, or you can admit the inevitable, and deal with what might appear to you as a mess.

The real answer is that making choices really isn't entirely the system manager's problem. There should be guidelines about archiving data (putting it on tape) and storing it offline. As a system manager, you would never willfully delete the users' data files, even if they appear to be useless or outdated. Deleting users' files will undermine your cause as a protector of their information. This is a sacred trust between yourself and the user community, remember the "habeas corpus"?. Users should be made aware of the guidelines early and often.

Many modern network operating systems have a provision for keeping ownership information for each file. In this way, you can use the networks' own (or third-party) reporting software to generate reports for the users based on time/date stamp information or ownership information for each file stored in the shared data area.

The Guest Account

Most systems are also configured with a second type of account. This is called the "Guest" account. Typically, it is a good idea to deactivate this account, because it serves no real purpose, other than to add ambiguity to your network. Generally, on a small network, there is no advantage to having anonymous accounts, as it opens a door to the system, one that cannot be readily audited or controlled.

Ordinary User Accounts

A good rule of thumb: Open just enough doors to let people do what they need to, but no more. Just as a user should not have need to change the settings on a server, they should also be limited in certain ways on their own computers.

On my network, the primary user for any client can change display settings, color and desktop, screen saver, and other normal settings. They cannot change the time, as this is set centrally at login by the network, and they cannot change the selection of mapped drives, although they can map additional drives if they choose.

I install a host of programs for all users, for basic tasks such as Internet access, email, and productivity. I also allow users to install and remove programs from the computer, if they so choose. While all users can change their screen savers and color schemes, I do not allow them to have services like email sending/receiving programs, FTP servers, or other odd servers running on their computers. In the heyday of peer-to-peer file sharing programs like "Napster", letting users run these services on their machines can have unpredictable results, and can work to welcome intruders and viruses to your system.

Here's an illustration of why control is important. Recently, we had an attack of an email based Macro virus. The virus was a modification on a well known virus called KLEZ. Somehow, this virus passed through our protection software on our email server. It then systematically damaged the antiviral software on each of the client machines. Finally, it homesteaded on our network, reaping havoc everywhere. The insidious part of this virus is something called an SMTP (Simple Mail Transfer Protocol) server. SMTP servers are usually a legitimate thing to have on a network, because they facilitate the sending of internet email. There should be only as many as you have set up for this purpose.

A great deal of the damage that KLEZ caused was because it set up illegitimate SMTP servers, which sent a lot of junk mail to everyone, basically clogging up our email program and network. At one point in this attack, we were receiving and sending over 20,000 messages each day. The attack lasted about 3 days until we were able to completely extricate it from our network.

Fortunately, we had previously locked down, using our firewall, any SMTP activity from anywhere other than our one mail server. I believe that the use of the firewall really prevented what would have been a multi-day network outage.

Private User Data

Users should not have any data stored locally on their own machines, and should have private space on your network. There should also be public space, so that people can share data without having to email attachments or run around the office with floppy disks. This allows you to maintain the entire data store, and to know that data is being backed up.

It is also my recommendation that you use a standard format for email and user accounts, and that they match each other. When I began my job, there were no discrete user accounts, just a guest account and an administrative account. The email addresses were individual, but were named things like FRJACK (for Fr. Jack) and SRROSE, for (Sr. Rose)—Well, I am working in a Catholic organization.

There was a dedicated volunteer running the system when I started, and I'm sure that this naming scheme made sense when they began. The problem was that as the system grew, these names had to be abandoned in favor of a scalable system. First names are not scalable, because when you hire the second Chuck or John or Sr. Rose, what do you do?

The first thing I did was to change this, and to make the naming scheme for both the network and email synonymous. I used the first letter of the first name, and the last name. A typical user account, for instance, for a Joe Brown would be JBROWN. If another Joe Brown would be hired, his middle name would be used, like JDBROWN, if his middle name was David. I would recommend staying away from numbered accounts, like JBROWN1, JBROWN2, et cetera. Names should be as personal as possible, because these are citizens of your system. Numbering systems inherently contradicts this and undermines morale.

Additionally, although you should set a default password for every account, There should be *no* accounts that have either the default password or no password. Make the system require that each user reset to a new password the first time that they log in to the system. Most networks do this by default.

A Few Words about Groups

Groups are lists of users who have accounts on your system. The same computer that maintains the user accounts on the network maintains groups. A popular idea amongst network software designers is to assign the group to have rights, rather than individual users. This can be a good idea, and can save you time when changing user rights. For instance, lets say you have a group called "ACCOUNTING". Accounting is a group of people, who amongst other things has access to the folder on the server called "Payroll Folder". You don't want everyone accessing the Secret Payroll folder, because then the payroll would no longer be a secret, would it. You assign the folder a permission, indicating that only people who are part of the group called "ACCOUNTING" can access this folder. The network administrator account is the only account that can add the people into the ACCOUNTING group (or any other group, for that matter), so the folder is safe.

Here's how the folder and group are set up:

FOLDER NAME: "PAYROLL FOLDER"
"ADMINISTRATOR"   (FULL CONTROL)
"ACCOUNTING"   (READ/WRITE ACCESS)
All Others   (NO ACCESS)

GROUP: ACCOUNTING
Jliberto: Joe Liberto, *Head Accountant*
Msmith: Mary Smith, *Associate Accountant*
Kmcnally: Kay McNally, *Part Time Accountant*

With this arrangement, The Accounting group has access to view the payroll folder, so does the administrator. All other users have no rights to the folder. If someone new joins the group, they need be added only to the group list, and they will have access to all of the resources accessible by that list.

Why is this any more advantageous or simple than just using the users' names to assign rights? In a situation where there are few users and only a few shared folders, there is no advantage. But, let's say there are ten folders, and thirty users. The users need be added only once to the group, then the group can be added ten times to the permissions for the folder. This equals forty actions, as opposed to 300 actions if you were using user names only. Because people need be added only once, the use of groups also limits the possibility of human error when making additions or changes to the system.

EMAIL ACCOUNTS

Electronic mail services are something of a double-edged sword. Email is a service that must be provided, although sometimes you'd like to avoid it. Like voice mail, and facsimile, it has become a staple of communication in modern businesses everywhere. It is very likely that you organization's impetus to put in a computer system in the first place was to have email accounts for the employees.

The key to managing email is the same as most other things, give the keys to those who need to have them. If each person in your organization needs email, give each his or her own account. Account sharing in the application of email is worse than in the realm of the network. Multiple people sharing an email account will no doubt cause confusion for your clients and coworkers.

Additionally, if a person does not need electronic mail services in the course of their job, these services should not necessarily be provided. While it is true that electronic mail is not usually more expensive to run with 10 accounts or 100 accounts, the cost in terms of productivity can be staggering.

I have worked in installations where people abused email. I saw one situation in particular where a person was involved in an amorous relationship with someone over the Internet. Daily, and constantly, they were exchanging messages. Ultimately, the access to email had to be discontinued, because it was seriously affecting the performance of the individual in question. This person worked as a telephone operator, and had little need for a computer having Internet access at all, much less an email account. It was in the earlier days of Internet, and the firm did not have a handle on who had access to such services.

In this day and age, even the smallest firms have control over these issues. As a network administrator, you should work with other supervisors to build a policy for email use. Does everyone get email access, well that's up to your organization. Make sure, though, that policies regarding email use, especially appropriate email use, are specified in the employee's manual for your organization. If no such manual exists in writing, make your own manual for technology issues.

TECHNOLOGY MANUALS AND WELCOME PACKETS

Technology is a great thing. Without structure it can destroy an organization. Unmanaged, unleashing technology on your organization is tantamount to placing cable television at the desk of everyone in your organization.

A policy manual that deals with technology issues can greatly ease the burden of having to explain the "rules" to everyone. You need to have rules, especially as your number of users grows. There are legal liabilities, especially in certain types of organizations that may involve children. You need to address issues such as appropriate things to send in electronic communications, appropriate use of technology, and the ramifications of violating these rules of integrity. The rules should be clear, and concise.

The system administrator, however, should not define any rules, unilaterally. The role of system administrator is not as a "policeman", or as a spy for the management in the organization. Human resources and business management people, as well as supervisors and senior level management will have more than enough in the way of ideas about what these policies and repercussions should be.

Welcome Packets

Your welcome packet is more than a rulebook, though. Welcome packets can include lots of useful information for the users of your network, such as a logical listing of services that are provided, and maybe some diagrams of how the network is set up. As opposed to going it alone, there may be a policy and procedure manual already available in your organization that you may be able to become a part of.

A section of your welcome packet might include a page for new employees. New employees need to know certain types of pertinent information. Amongst this information might be:

- The Address of the business

- The main telephone number

- The main fax number

- The website address

- The User's username, and email address

No matter how small your organization, you should also have a typewritten, up-to-date telephone and email address listing, including the correct spellings and titles for everyone there. This can help to alleviate a great deal of stress for a new employee.

Technology Use Covenant

A technology use covenant is the first piece of documentation that should be drafted. This document is your contract with your users. Any user wishing, or needing a user account should read through, sign and date the packet. This should apply to anyone using your organization's system, whether they be a volunteer or paid staff. In a small organization, your users should be involved in defining this document.

There are certain sections that may be defined by outside interests, the laws of the state, for instance.

<u>Most documents will have the following sections:</u>

- Introduction, with version and date information of the packet

- Overview of Computer "assets" and Network Services Channels

  - Description of Services

    - Internet and the World Wide Web

    - Computer-based online services

    - Electronic mail and messaging systems

    - Electronic bulletin board systems

  - Description of Assets

    - Physical Resources

    - Employee information

    - Employee benefits and insurance information

    - Databases and the information contained therein

    - Computer and network access codes and similar or related information

    - Contractual and proprietary information

    - Research projects and all related information connected with research efforts

    - Other confidential or proprietary information that has not been made available to the general public.

- Principles of Acceptable Use

  - Use for Positive, Ethical purposes in congruence with your organization

- Use for Educational purposes
- Use for support of organizational activities
- Use in accordance with other organizational Policies and procedures

- Principles of Privacy

  - Definition of privacy
  - Privacy of Employees', Volunteers' information
  - Privacy of Organizations' information
  - Privacy of clients'/people-base's information

- Legalities

  - The laws of the state and federal government
  - Software Piracy statement
  - Ethical Use of information statement

- Safety

  - Statement regarding safety and security of organizational volunteers, and employees
  - Statement regarding safety of organization from electronic vandalism
  - Statement regarding willful destruction of property

- Safety and Protection of Children

  - In this day and age, this section may require a great deal of planning and discussion.
  - Basically, this section is in regard to child protection and pornography issues.

- Unacceptable Use

  - For activities unrelated to official assignments and/or job responsibilities, except incidental personal use in compliance with this Policy

  - For any illegal purpose

  - To transmit threatening, obscene or harassing materials or correspondence

  - For unauthorized distribution of organizational data or information

  - To interfere with or disrupt network Users, services or equipment

  - For private purposes, whether for-profit or non-profit, such as marketing or business transactions unrelated to organizational duties

  - For any activity related to political causes

  - To advocate religious or political beliefs or practices contrary to organizational values

  - For private advertising of products or services

  - For any activity meant to foster personal gain

  - Revealing or publicizing proprietary or confidential information

  - Representing opinions as those of your organization

  - Uploading or downloading commercial software in violation of its copyright

  - Downloading any software or electronic files without reasonable virus protection measures in place

  - Intentionally interfering with the normal operation of the organization's Internet gateway.

- Enforcement and response to violations of this policy

  - Specific courses of action, may refer to an umbrella personnel policy

- Acknowledgement and signature page with version and date information of the packet *(which should be separate from the packet itself)*

GUESTS OR VOLUNTEERS

While many things may be the same, there is one key difference between for-profit and non-profit organizations: Volunteers and Guests. Very few for-profit companies have volunteers and others not under the direct control or employ of the organization involved in its operations.

Volunteers are a crucial resource to any non-profit organization. They bring a fervor and energy to an organization, one that is rarely found amongst the paid staff. However, as a network administrator in a non-profit, I am always dealing with questions and issues surrounding volunteers using our resources and services.

Why would volunteers tend to have a need for your technology? In our installation, we find that there are two answers to this question. Either they are producing some kind of product, which is based on software which they do not personally own (eg. a web page, or a brochure) and are using your equipment to complete this project.

Alternatively, the volunteer is checking or working with information stored on the system, like e-mail or a database that would not necessarily be available to them outside of the network.

These are both good reasons for volunteers to use your equipment. Part of the overall management of volunteer efforts in this area involves communication and group involvement. While sometimes people work better on their own timeframes and in their own locations, this is not always feasible. Especially in the two situations above, you may not be willing because of security, or able because of licensure issues to give them the resources that they would need at home.

Volunteers need their own set of management rules. Like contractors, volunteers with expertise can make a big difference in quality when implementing specific projects. Keep in mind, that by nature, they cannot be dealt with like employees. Volunteers are working for your organization out of a sense of purpose or community involvement, and not for financial security or institutional power.

All too often, volunteers tend to be out of the loop on a lot of managerial and policy issues. Most of the time, volunteers are working on narrow-scope projects, usually at non-standard hours, and in many social ways, parallel to the rest of the 9 to 5 crowd. In our organization, volunteers tend also to get a lot of information third hand, and are not usually part of training and education efforts made available to the rest of the [paid] staff.

There are a few ways to remedy communication inadequacies with volunteers. One is by the same kind of regular communication that is already in place, usually by the employee associated with the committee or volunteer effort. This is the ideal situation, but not always realistic. I have found that in my organization, certain of the paid staff people are not interested in the policy issues surrounding technology, especially if these issues do not directly relate to something that they do in their day-to-day activities. These policies, or changes, rarely make it reliably to the committees and volunteers. Many of the important policy changes never make it to the tertiary level of the volunteer groups.

The best resolution may be through your *technology committee*. For more on this, see the section later in this book, pertaining to technology committees. Email communication or other traditional means of communication can be helpful. Keep in mind that e-mail is a heavily saturated medium for most people. While occasional use of e-mail for policy disbursement is probably moderately effective, you will probably find that its effectiveness is inversely proportionate to the amount of policy information that you disburse.

Outside of policy management, there is the issue of managing volunteers' logistical resources, like user accounts. With volunteers more so than with your employees, appropriate rights assignment and user control is critical.

Most of your volunteers are honest, hardworking people. With the right encouragement and goal development, volunteers can produce equal, possibly even greater quality of work than the people working in your organization on a day to day basis. With structure, and encouragement, your volunteers will eventually become long-standing members of your organization, the value of these people cannot be measured.

This being the case, it is important to realize that, especially with volunteers working after hours, maintaining system integrity and security is critical. Unsupervised access to your resources is a temptation for anyone, a temptation that may be too great for some people. These resources may be information, access to your electronic devices, printer, photocopiers, or access to your software. Just as your organization protects the facilities, you must protect the data and system resources, whether or not they are physical assets. Many volunteer organizations are socially active, keeping records on their target community or in some cases on their clients. In smaller communities, there is always the threat of people trying to excavate information about each other. You, as a protector of the information, must be aware of this potential security risk, and do what you can to prevent it from happening.

The first step is to analyze the project that the volunteer is to do. Give the volunteer specific goals and objectives for their project. Write these goals down, and use this list to determine what the necessary programs and data will be.

Once you have determined this information, make their user accounts comply with their needs, no more or less. In addition, make program icons and folder icons as readily accessible as possible, so that the need to "explore" is reduced. As with any employee, try to make the project engaging and structured enough as to maintain interest and limit unrestricted time on your system.

As potentially unnerving as unsupervised access to your sensitive information is the same regarding your Internet and email connections. Do your volunteers need to access the Internet? If so, you may wish to employ some form of filtering software, or barring this, you may wish to work with your volunteers to schedule times and locations where these projects can be completed with supervision.

Important with all users as well as volunteers are the logistics of time restriction and account expiration. Contractors as well as volunteers should have time-limiting controls programmed into their user accounts. This allows the system to dictate the times when these types of users can use the network. Volunteer accounts, like other user accounts, should be deactivated when these users leave the organization.

This is not so much an issue with volunteers specifically, as it is a security issue involving all of your users. There was a famous novel about network security written in the 1980s. The book was titled "*The cuckoo's egg*". If you haven't read this book, I highly recommend it. Its story is timeless and the writing style engaging. The weak link of their system laid in the fact that they were an educational research institution in the earlier, headier days of computing. The point of this book was that through lax user security, that people outside of the system were able to take advantage of their network. Ironically, it was through the auditing of their accounting system that they realized the breach and caught the intruder. Today's smaller and more connected systems, this threat is as real as ever. Unfortunately, most networks do not use similar tools, because the idea of "accounting for system use" is no longer in vogue, and system managers might not catch such a problem.

Volunteers and Training

Training of new people and continuing education of existing parts of the organization is critical to maintaining your system. Volunteers are no different. Training of volunteers is a twofold process of ensuring technical ability and maintaining clearly defined goals and guidelines about nature of the project.

Depending on the nature of the project, training may be very simple, such as teaching someone how to open a database and do data entry or verify information. Some of your higher-level volunteers may want to be trained to do things such as run their committee's portions of the website, or make brochures, or other types of projects.

My opinion on this is that you may wish to send these higher-level people to professional training programs or seminars. This may be a complex decision for your organization. There exists a certain stigma about sending volunteers to training classes. Granted, it requires judgment on your part because volunteers have no inherent responsibility to your organization. Your volunteer may choose to get training outside of the scope of your organization. If you find yourself In this position, give them a copy of the goals of the project to balance against the outline of a potential class. If you have kept up on training resources, you may be able to recommend some options.

Alternatively, you may be in the position of needing to send someone to training classes. If you choose to sponsor a person for a certain program, you may wish to sit with the volunteer, one-on-one, and discuss their mid-or long-term plans as they relate to your organization. As hundreds or thousands of dollars can be potentially spent on a training class, getting a verbal agreement for some level of commitment is a good idea.

Successful training also involves use of the skills within a reasonable time after completion of a course. Training, especially of a technical nature will only stick if it is applied.

DOCUMENTATION

Whether considering volunteers, paid staff, or contractors, documenting your system is crucial to communicating its layout. Documentation projects far exceed the scope of technology installations in an organization. All facilities management or line-of-business management positions require the keeping of paperwork and plans having to do with their respective areas of responsibility. Policies and plans are codified, and must be organized for future reference. For technology departments especially, organizing documents is crucial. If there is equipment failure, a system failure or a change in personnel, adequate documentation can save your organization a great deal of time and money. There are numerous forms of documentation. Keeping folders containing service contracts, warrantees, and instructions for equipment should begin at the outset of the project.

Think of documentation as an ever-increasing library of files. There are several types of information that should become part of the technology department's documentation library.

- Receipts and service guarantees

- Purchase orders and Service Contracts

- RFQs and Contracts

- As-built (post installation) documentation

- Internal Memos and budgets

- Policy Information

- Training Information

- Inventories of equipment

- A Graphical Plan Documenting the Overall Infrastructure

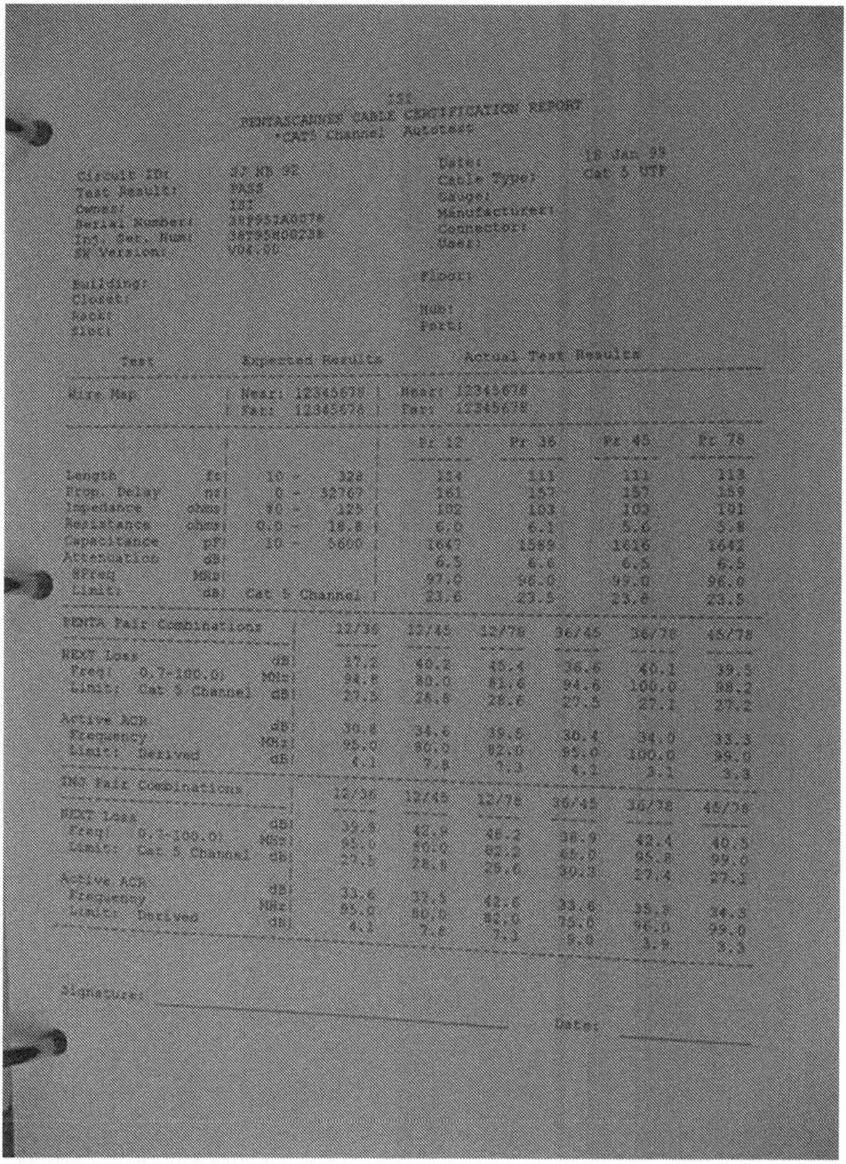

ISI
PENTASCANNER CABLE CERTIFICATION REPORT
*CAT5 Channel Autotest

Circuit ID:	2J KB 92	Date:	18 Jan 99
Test Result:	PASS	Cable Type:	Cat 5 UTP
Owner:	ISI	Gauge:	
Serial Number:	38F95IA0078	Manufacturer:	
Inj. Ser. Num:	38T95H00238	Connector:	
SW Version:	V04.00	User:	
Building:		Floor:	
Closet:			
Rack:		Hub:	
Slot:		Port:	

Test		Expected Results	Actual Test Results			
Wire Map		Near: 12345678 / Far: 12345678	Near: 12345678 / Far: 12345678			
			Pr 12	Pr 36	Pr 45	Pr 78
Length	ft	10 - 328	114	111	111	113
Prop. Delay	ns	0 - 32767	161	157	157	159
Impedance	ohms	80 - 125	102	103	103	101
Resistance	ohms	0.5 - 18.8	6.0	6.1	5.6	5.8
Capacitance	pF	10 - 5600	1647	1569	1616	1642
Attenuation	dB		6.5	6.6	6.5	6.5
@Freq	MHz		97.0	96.0	99.0	96.0
Limit:	dB	Cat 5 Channel	23.6	23.5	23.8	23.5

PENTA Pair Combinations		12/36	12/45	12/78	36/45	36/78	45/78
NEXT Loss	dB	27.2	40.2	45.4	36.6	40.1	39.5
Freq: 0.7-100.0	MHz	94.8	80.0	81.6	94.6	100.0	98.2
Limit: Cat 5 Channel	dB	27.5	28.8	28.6	27.5	27.1	27.2
Active ACR	dB	30.8	34.6	39.5	30.4	34.0	33.3
Frequency	MHz	95.0	80.0	82.0	95.0	100.0	99.0
Limit: Derived	dB	4.1	7.8	7.3	4.2	3.1	3.3

INJ Pair Combinations		12/36	12/45	12/78	36/45	36/78	45/78
NEXT Loss	dB	39.9	42.9	46.2	38.9	42.4	40.5
Freq: 0.7-100.0	MHz	95.0	80.0	82.2	86.0	95.8	99.0
Limit: Cat 5 Channel	dB	27.5	28.8	28.6	30.3	27.4	27.2
Active ACR	dB	33.6	37.5	42.6	33.6	35.8	34.5
Frequency	MHz	95.0	80.0	82.0	75.0	96.0	99.0
Limit: Derived	dB	4.1	7.8	7.3	9.0	3.9	3.3

Signature: __________ Date: __________

FIGURE 2.25: CABLE CERTIFICATION DOCUMENT

An Example of Documentation in the form of a Cable Certification Sheet. This sheet certifies that one jack is compliant with the

EIA/TIA Cat5 standard. It is not necessary that you understand all of the technical aspects of this document. It is important, however, that the network is certified, indicating that each of the jacks has "Passed". Verify that you have received a binder or book containing these specification sheets, one for each jack installed on your network.

Product Documentation

The first four goals in the previous list can readily be achieved. Keep a file cabinet with files pertaining to each isolated installation or upgrade, or by manufacturer or vendor. I have found that whenever I am referring to this type of documentation, I need some piece of information about a particular product or service. I have folders for each piece of equipment, and keep these folders on file until the project or equipment is retired or supplanted by new technologies.

An example of this is our recent calendar server project. The calendar server project was two parts, so there are two files. The first file had to do with getting quotes and information, this file would contain:

- Internal Memos and Explanation of our Calendaring Process

- The Needs Analysis Handout

- The Needs Analysis results

- My notes from the meetings with our staff

- (Print-outs of] Email correspondence about this project

- The RFQ

- The proposals and product literature from the vendors

The second file has to do with the actual product and vendor information.

- The Contract and scope of work document

- The invoice and receipt for payment

- The software licenses

- Copies of the product literature from the vendor

- Copies of the server documentation and warrantee

- Letters from the vendor about the project

- Time frame Schedules, and my notes

- Memos to the staff about implementation

- Internally produced Training Documents

In addition, a box is kept in the system room with installation and instruction manuals for both the server and software, and the diskettes and related paraphernalia.

A different type of installation documentation may involve the user's desktop computers. If your organization purchases PCs in quantity, you may not need to keep multiple copies of the product literature.

In our organization's situation, storage space for such documentation is too limited to keep numerous duplicative copies of books and documents. I keep a representative binder for each "generation" of computer purchase. This binder includes the copies of the receipts, one or two copies of the computer documentation manuals, any applicable licenses for all of the computers, warrantees for all of the computers and monitors, and driver diskettes. This way, all of the documentation for a particular kind of computer is readily accessible.

Many pieces of costly equipment, even after retirement, are kept in storage for possible future use. At the time when a server or other equipment is retired, I will continue to maintain the documentation. I will, however, put a note in the file regarding the current location of the equipment in storage. This would help a subsequent network administrator person to locate a piece of retired equipment for future use.

Budgets and Memos

In regards to internal communications, budgets, and other pertinent documents generated by the technology department, we keep an electronic copy of all of these files on the network server. In this way, all of the memos and source documents are kept at ready access for future use.

In the same way, policy information, drafts of policies, and source documents for policy documents outside the realm of our technology department are kept on electronic file. The final policy documents in our organization become a part of a larger binder of Policy documents, which have to do with many other things besides technology.

In my system, I have kept an electronic file folder for each "budget year" since I have been in this position. Each folder contains all of the memos, internal documents, and the proposed budget for each year.

Training Documentation

Training documents differ in nature from receipts or internal communication. Typically, documenting training is more difficult. Personally, I do not maintain documentation for contracted training for individuals. For instance, if I send on of our administrative staff to training off-site, I do not keep records on this training, for information other than the invoice, and anything which may have been sent to me from the company. The business office's function in most organizations is to maintain copies of all the invoices and records of that nature.

Internally administered or custom training is a different situation. Usually, in the course of setting up this system, we have brought in an outside trainer to do group training for our employees. In this case, a binder is kept for each training class, or alternatively, a representative piece from their handouts is kept on file.

Inventories

Maintaining up-to-date inventories is paramount in any organization. When speaking to inventories, I am usually referring to hardware and software products that have been purchased for the organization.

There is no one solution for maintaining inventories. Many business offices keep inventories for accounting and insurance purposes. If you organization does not do this, it should consider implementing one. Insurance inventories should include basic information about the age of technology equipment, and serial numbers. In the event of a robbery or other loss due to destruction, the insurance company calculates approximated values. These values are usually based on the age of the equipment rather than on their replacement value.

Robberies are a likely eventuality in any sizable organization. Even in our facilities, which are monitored by after-hours security people, we have had some equipment stolen right from people's desks. Last year, for instance, a laptop computer was stolen off of the desk of an associate. The shock to us was that it occurred during business hours. The computer was stolen while we were all in a staff meeting and out of our offices!

A capital equipment inventory would include all of the computers, monitors, servers, network equipment, telephone equipment, and other A/V hardware. It would also include any physical software packages, books, or other valuable items which the technology system may have. A simple database can be devised to keep track of this equipment.

LOCAL ID NUMBER: 0034
SERIAL ID NUMBER: 004-3002325
DESCRIPTION: PENTIUM PC/133 MHZ
PURCHASE DATE: 02-21-2000
LOCATION: 1ST FLOOR MARY'S OFFICE

A possibly more useful inventory to you may be of the specifications variety. Inventories of this type are devised so that you can begin to analyze any adverse effects of an upgrade, or analyze the equipment to work on equipment replacement schedules. A simple database of your computer equipment can be created for this purpose.

In our organization, this type of databases keeps track, for each machine, of the following Information:

- The computer's Network Name—The name given to the machine to identify it on the network

- The computer's Location, specified here by building code and room code

- The Brand of the computer

- The Speed of the Processor

- The amount of RAM memory in the computer

- The hard disk drive size

- The role of computer, or rather, its purpose as a client, or server

- The Operating System

- The Screen size

- The computer's IP number, as assigned by the network manager

- The physical data jack number to which the computer is attached

Whether planning for upgrades of some type, or performing regular maintenance to computers, it is a good idea to keep a comprehensive, up-to-date inventory of the equipment on the network. Keep in mind, however, that the more varied information kept in each record (for each machine), the more manual updating you will have to do in the course of maintaining the inventory records.

THE WEBSITE

The product of a tool manufacturer may be drills, saws and hammers. The goal of this manufacturer is to make a profit, and thus remain in business. The product of a school is an educated individual, the goal to remain competitive amongst other similar-caliber institutions. Though it is not drills or hammers, It is likely that you have goals and products as well. The product of your not-for-profit may be some form of activism or social justice, such as a healthier homeless population or well-adjusted, cared for children. As most non-profit organizations are somewhat altruistic, it is likely that the goal of your organization is to help improve the life of some group of people. Ask yourself, What is the product of your non-profit organization?

One important way to meet your institutional goals is through effective communication with others. Others may be people inside or outside of your organization. Getting your message out to people is paramount to success in an organization which, by nature, is about helping other people.

In our organization, the primary means of communication have been traditionally through paper mailings, our weekly bulletin, and our monthly newsletter. People are beginning to embrace the website, and to realize that used correctly, it can be a far more powerful medium.

Our organization has had a website since the mid-1990s, well before I took over the management of the system. Its purpose started out minimal, mainly because one of the volunteers suggested a web presence. It was simple in design, gave directions to our facilities, and covered all of our groups and departments, if no more than to provide contact information. While some of our groups began using the site more actively, posting pictures, others have never really embraced the site. Our technology committee is in the process of bringing people information about the technology, and asking that they consider its application when carrying out their various missions.

Like many aspects of technology management, work on your organization's website will grow exponentially as time progresses. It may start out simple, but as one or two groups begin putting time and effort, other people and groups will want to become a part.

As a technology manager, it is unlikely that you will want to manage the entire site. In all but the smallest organizations, the website requires more management than one person, a person for whom duties are varied, can handle on their own.

The Pyramid of Responsibility

Our organization has adopted a top-down structure to the site, which allows our people to manage their own content, while not compromising the control that we need to maintain internally.

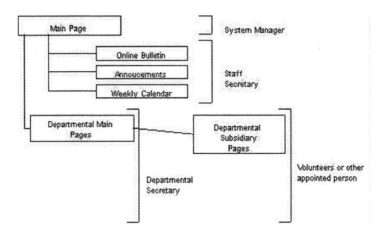

FIGURE 2.26: THE PYRAMID OF RESPONSIBILITY

In our management scheme, the system administrator maintains the central page. This allows a complete control over the look and feel of the title page, and the addition or deletion of any links to subsidiary pages or content.

Linked to the main page, is an online bulletin, special announcements and notices, and a weekly calendar. These pages are maintained by the staff secretary, who has been professionally trained by a third-party training organization in the operation of the website management software.

Linked to the main page are all of the departmental pages for our religious education center, our school, and our other specialized groups. The look and feel (color and typefaces) of the "second level" main pages are defined by myself. Aside from these guidelines, the content is defined by the department secretary for each of these departments, in accordance with what each department wants on their page. Subsidiary to each "second level" page, the work of maintenance is defined by each department. For some groups, volunteers do the content, for others, the secretary does the updates. Because responsibility over the content always belongs to our organization, regardless of who creates it, the actual updates are always managed by employees.

Web site contractors and knowledgeable volunteers can make a huge positive difference in the overall quality of your website. This is especially true in the look and feel, and in other "soft" issues, such as usability. Typically, individuals use consumer-grade website development packages. While these programs are easier to use, and are usually less expensive, they are philosophically a lot like using a home video camera.

Marketable home video equipment was devised about twenty-five years ago. The concept was as a replacement of an existing standard, the 8mm film movie camera. Video cameras were designed to be:

- Small and as lightweight as possible

- Easy enough to use for the ordinary individual

- Safe for ordinary people to use

- Economical to purchase and operate

186 Technology and Purpose

- Able to operate on batteries

Notice that the word "quality", or "broadcast like" are not in this list of features. The electronics industry knew that people would rather sacrifice quality for ease of use and economy. Certainly, with the video camera you can tape things, but there is a marked difference between what you can do with a video camera and what a television studio and a production company can do. While some of these kinds of issues may not concern your organization, you may wish to have a dedicated website programmer help you to develop a good looking, usable site.

A rule of thumb in smaller organizations is to shy away from "glitzy" extras, when programming any web site, you should make a list containing the "lowest common denominators" which will work adequately with your site.

When making this list, you will want to look at your community of people, and the groups of people who will be likely to access your site. Are these computer literate people, or are they people who may be new to technology. The site is not likely to be a technology oriented site, in terms of content, so you may not want it to be too feature-rich, or confusing to navigate.

- What kinds of users are you trying to attract?

- What size of display/Display Resolution do I need to see the entire site on the screen?

- How fast of an Internet connection do I need to use the site?

- How large in size, measured in Kilobytes, should the biggest picture allowed on the site be? Excessive numbers of pictures, or large pictures, will create a delay, and thus hinder usability.

The chief problem that plagues in web sites today is that the site designers do not take into account the types of users who are likely to visit. The visitor may not have a 17" display monitor, or even a brand-new computer. In today's market, it is very likely that a home or small-business user is using a dial-up connection. Your site design should reflect this by making intelligent use of feature-rich options.

Any kind of scripting or page programming on a front page of a website, I think, is a bad idea. Many older browsers will not support, or will react in unexpected ways, to scripting in web pages. More and more corporate users have options disabled to prevent attacks from the outside, and potential portals to virus problems. Scripting should be viewed as an option which, unless necessary for the sake of functionality, is to be avoided.

To accommodate lower-bandwidth users, your website should have a link to a "low bandwidth" version of key pages. This version of the site may be less feature-rich than the regular site, but will allow people with slow connections to bypass large graphic files, without having to stop their browsers from downloading these files manually.

Other visitors to your site may include the visually impaired. Making your site accessible to the visually impaired visitor is like making your buildings handicap accessible to those in wheelchairs. The visually handicapped make up a surprisingly large group of Internet browsers. The medium, when set up correctly, can be very friendly. Visually impaired users typically use some kind of device or software, which converts the written text (but not graphics), to a computer-generated spoken voice. By listening to the voice, these users can navigate your website, or access pertinent information, such as your address or telephone or fax numbers. To accommodate this group of people when programming your site, there should be a link to a text-only version of the web page. Alternatively, sites can be programmed to automatically redirect a user to a text-only version of the website, if such a browser is being used. This may be a more elegant, or preferable, way to handle the site.

Integration

Some sites will be more information-rich than others. Many web sites will link to other sites of similar interest, or for beneficial resources.

One popular option right now is to coordinate the local content, that being your published information, with local files on your network. Sites can be set up which work with databases to allow data-entry from the website. The web server running the site then send this information to other types of files, like Database files, for instance.

There is a great deal of benefit to be had from an interactive system. Interactive web pages can improve the accurate and expedient gathering of information. In addition, if your organization conducts mailings throughout the year, this form of data gathering can supplement these mailings, and allow you to get more saturation and a better rate of response.

Web pages should always supplement, but not replace, mailings and bulletins. The problem with technology is that it is readily creating divergent groups of technology elitists and impoverished. Like other socioeconomic or cultural movements, there are people who for many reasons, will not, or cannot make the move to technology. In your organization, it is unlikely that you will want to shut the door to these people. At our stage of the game, we are trying to broaden the scope in which to get people's attention and support. Making information web-based only, or looking at the web as a way to save money is not always a good move.

FALL: PRODUCTION

o o o o o o o o o o o o o o o o o o o o o o o o o o o o o o

"May your life be like good wine. Tasty, sharp, and clear. And like good wine, may it improve with every passing year."

—*An Italian Blessing*

[The network contractor] delivered the servers and hubs today. I couldn't find a table, and there are none in the system room, so I borrowed a long folding table from the meeting room. The equipment sits on the floor, and they are connecting things together. The room is temporarily a mess of wires, cardboard packing crates, and plastic blinking boxes. They are asking questions about the network setup. What is the network's name going to be. I think for a minute, and respond. And so it is born and christened, "SJS"—St. Joseph's System. They ask me what I want to call the two servers, the first two servers.. I say to the technician, let's name them after the two rival columns of power in any academic culture. We'll call them "Academic" and "Administration".

September 1999

REVIEW AND SUPPORT

It is a time of harvest and production. Lofty goals and fantasy will begin to give way to reality and the tests of time. This is the third and longest phase of the life of a network. The contractors and installers are gone. The Bills have been paid. New people may come, and old ones go in your organization, but the system will remain.

Your systems will fall under the scrutiny and test of the hardest judges of all: The Users. The system will get older in this phase, and will need a little help along the way.

The third phase really starts after the completion of installation. Once the network has been installed, it is important to make a list of priorities for the continuation of service. Undeniably, the most expensive resource is the human resource. While you may stock parts, extra systems, et cetera, there is likely only one or two of you, and a host of contract people. Because especially when the responsibility begins to require more than one support person can handle, there needs to be prioritization of what gets fixed first.

1. INFRASTRUCTURE OUTAGES

2. VIRUS-ATTACK RESPONSE

3. BACKUP FAILURE

4. INDIVIDUAL USER OUTAGES—CLERICAL

5. INDIVIDUAL USER OUTAGES—EXECUTIVE

6. CONVENIENCE ITEMS

Any problems that occur at whole-system level should always take priority over an individual's problems. Responding to virus attacks and attacks of the network, and any backup-system problems should also take top priority. Your individual users may not agree with this philosophy. Nonetheless, it is crucial keep network-level services running, so that the majority of your users can get their work accomplished. These policies should be codified before problems occur, so that the "Squeaky Wheels" amongst your staff can have an explanation of the rationale. For more information about problem resolution, see the section about work-orders and problem response later in this chapter.

Looking at the system as a whole, you may wish to compile a bullet point list containing your priorities.

1. Safety of Data

2. Protection of network against outside attack

3. Availability of Key network services (eg. Routers, hubs)

4. Protection of the key physical hardware resources (eg. Routers, hubs, servers)

5. Availability of system services (eg. Servers)

6. Availability and protection of client hardware (eg. Client machines)

7. Availability for individual clerical users (eg. Secretaries who depend on the data system to perform the majority of their jobs)

8. Availability for individual occasional users (eg. Users who do not depend on technology to perform their jobs or those whom have readily accessible backup systems—eg. A notebook computer or equivalent)

9. Acceptable levels of usability for system services

10. Regular Maintenance items

11. Organization of system documentation and resources

There is one unchanging reality in technology. In general (outside of the realm of your own organization] more data communications systems are being installed in business ventures. As this occurs, the general level of reliance on technology as a function of business continuation will increase accordingly. The need for timely response to problems and development of your systems will only increase as the amount of reliance on this technology increases. To counter-act future turmoil, your organization needs to respond proactively to service and development of technology.

A position of systems administrator or technology coordinator can help to fix for this. An administrative position is usually created as a response to a need for support and administration for a system. Usually this first and foremost includes a data network. Depending on the organization, this may also include infrastructure wiring, telecommunications systems, video distribution systems, and/or audiovisual systems. The position may be responsible for administration, repair, or support of these systems. It may include system rooms, offices, education or training centers, or other special-purpose data installations. Your system manager may have other responsibilities in addition to the management of the system, like business management or office management.

There is a slow transition from the implementation phase to the support and review phases. As this occurs, your needs will change, from the hectic environment of building, to a more stable, consistent environment of support and maintenance. This is the idea, anyway.

Systems rarely work this way. Much more typically, the system moves into a more response-oriented mode, but the building and changing is not complete. As time progresses, and your systems have been installed, you will have been in the role of administrator for much longer. Certain unwritten, albeit real needs will likely surface. The system moves into a review mode, which spurns additional planning for both new and renovation projects.

You may look back, and realize that you moved quickly to get certain types of service available. After the stability of the system is achieved, you can go back and redo these services, giving them much more attention than may have been previously possible.

Support and the regular maintenance of your system is a very real need, as well. Continued maintenance will make the difference between a working, reliable system and large bills from computer contractors. Organizations can handle this issue in different ways. Some handle this from the outset, by contracting with a particular vendor who provides the services indefinitely. Alternatively, some organizations choose to hire a full, or part-time system administrator to maintain the systems.

In my estimation, unless your system comprises more than 30 or so computers, and you desire real network structure and services, the full-type network administrator is probably not necessary. There are many freelance system managers who operate businesses that provide system consulting and repair services. Outsourcing your maintenance, while more expensive per-hour, is actually less expensive, because factors such as insurance benefits and vacation time can be eliminated.

Outsourcing your planning or management is a different matter. Many system consultants will take this opportunity if it should become available. Sometimes, system planning is very casual. The consultant will come to repair or upgrade something, and will get into a discussion with the on-site person. This may lead to a new system or set of services that the consultant can provide. This can be a good idea, as the consultant likely has more experience in the field. This can also lead to the consultant driving the network strategy, which can be less than ideal for the organization. Too many organizations put excessive resources into unnecessary plans, many of which are devised by consultants, and not by real needs of the people in the organization.

TECHNOLOGY BUDGETS: THE BEST TOOL FOR SUPPORT

In many technology installations, too much of the initial funding is spent on implementing fancy systems. There must be a contingency from the first day to maintain the installation.

Depending on your avenues for service, this maintenance may be in the form of service contracts, local vendors, volunteers, or even an on-site system manager.

Your system's budget should take into account the amount of equipment that you are supporting.

Below is a sample of our technology budget for maintenance and repair issues. Our budget is done in two parts, one for both ordinary and extraordinary repairs and maintenance, and another for "capital projects". Our repair and maintenance budget is included below because budgeting for repairs is of key importance to the successful operation of a network. Because the dollar amounts are not important for this exercise, I have decided to show percentages in lieu of dollar values for this example. You may realize as we have that there are certain aspects of the system whose cost percentage will increases with time. Repairs and maintenance to client computers is a good example.

We have added many computers since we started the implementation phase. Donated equipment, purchased equipment, and upgrades have made the system grow about 1/3 in size larger than was originally designed.

There is a trickle-down effect that appears when adding equipment to a system. When adding computers, you must also look at the maintenance costs per machine, effects on licensing, effects on printing, and the effects on overall network performance. Additions of equipment may cause the network to require upgrades which were not foreseen when the decision was made to add to the system.

There are several sections of our budget. These sections are Capital Expenditures, Training, and Ordinary/Extraordinary repairs and maintenance.

Capital expenditures deals with requests to purchase new systems and services. It is based on planning for the upcoming year. Our management committees and management people look over this section, to assess the feasibility and cost of these projects

The section regarding Training involves all costs related to training individuals. This includes both the network support people, and the end-users of the system.

The final section, reproduced here, is the repairs and maintenance portion. This section deals with service contracts, repairs, and the like.

A Sample Budget

COMPUTING & TELECOMMUNICATIONS

ANNUAL BUDGET
For Maintenance Expenses
2002–2003
(SHOWING PERCENTAGES)

I. CONTRACT ISSUES

=-=-=-=-=-=-=-=-=-

A. DIGITAL TELEPHONE MAINTENANCE CONTRACT   3.5%
Maintenance agreement for our PBX system

B. TELEPHONE LINES MAINTENANCE AGREEMENT   2.5%

                                     6%

II. OPERATING EXPENSES

=-=-=-=-=-=-=-=-=-=-

A. HP LASERJET MAINTENANCE   6%
Per-unit maintenance for our hp laserjet printers

B. ANNUAL FILE SERVER VIRUS UPDATES   1%
Covers virus protection updates for network servers

C. PER-MACHINE COST TO COVER REPAIRS   23%
Covers repairs of existing end-user equipment

D. NETWORK DIAGNOSTIC EQUIPMENT AND TOOLS   1.55%
Covers cost of tools and equipment for repair/diagnosis

E. ALARM SYSTEM AND AUDIO SYSTEM REPAIRS   1.55%
Covers expenses outside of the contract

F. CONTRACTOR CONTINGENCY EXPENSE   12.6%
Covers expenses accrued if a backup contractor needs to be brought in
to do computer-related services

G. AUDIOVISUAL EQUIPMENT CONTINGENCY EXPENSE   .63%
EG. Replacement projection bulb for overhead viewer

                                                          45.93%

III. TRAINING/MAINTENANCE

=-=-=-=-=-=

A. PURCHASE MATERIALS FOR TRAINING OF VOLUNTEERS OR   8.68%
THOSE WHO PROVIDE SUPPORT SERVICE
Covers cost of providing training classes/materials

                                                          8.68%

IV. SERVICES

=-=-=-=-=-=-

A. INTERNET & BACKUP INTERNET CONNECTIONS   8.3%
Internet/web/email connection

B. TELEPHONE CONNECTION BILLS   8.3%
Covers all telephone service at parish/school

C. DNS NAME HOSTING COST   .07%

                                                          16.67%

V. PROACTIVE MAINTENANCE ITEMS

=-=-=-=-=-=-=-=-=

A. 10 PERSONAL COMPUTERS   11.1%
To replace old equipment in offices which is still in use

B. VIDEO PROJECTOR BULBS   6.3%
Covers cost to install video projector system in parish

C. TONER CARTRIDGES AND PRINTER CLEANING   5.0%
Covers ink costs

22.4%

BUDGET TOTAL FY2002   99.68%
(.32 variance due to rounding)

SYSTEM ADMINISTRATORS AND TECHNOLOGY COORDINATORS

This category envelops a great variety of people. The hired "System Administrator" may be range from a professional system manager, or an office manager who has been moved to this position, or even in the case of the part-time position, a student who has a talent for technology, and an interest in making some extra money. A good system administrator is probably a younger individual, possibly someone out of a MIS, Computer Science, or similar collegiate program.

Technology Coordinators are a different type of System Administrators, as they tend to be primarily individuals with experience in management. You will probably find that candidate pool for this position may include older individuals, the retired, or those interested in a part-time job, or a second career. With the recent events in industry, there are many management-oriented individuals who are good candidates for this type of position.

The key difference between a System Administrator and a Technology Coordinator is that the System Administrator position is likely to attract a more hands-on type of person. System Administrators tend to be more hardware or software oriented, may have programming or systems experience. They will be more involved in the nuts-and-bolts operation of a network.

Technology Coordinators may also be the business manager in your organization. The Technology Coordinator may have other responsibilities, such as managing budgets, or service contracts for other aspects of the organization. The technology coordinator position marries well with the business manager's position, because they share many of the same responsibilities.

Obviously, your funding, size and scope will define your technology plan and ultimately drive the levels or number of individuals necessary in the position. Although more costly, it may behoove your organization to hire a professional systems person, especially if that person is to have decision-making authority.

While going through your process of defining a position, you and your management people may wish to define a job-description for a technology coordinator or system administrator in your organization:

General Responsibilities of a system-administrator may include:

- Support desktop applications

- Support network services and applications

- Provide network, software, hardware and telecommunication administration—including long term planning of the system

- Coordinate and maintain Audiovisual System

- Build Automation Systems

- Coordinator of Website

- Liaison for a Technology Committee

- Develop on-going user training programs

- Coordinate with Business and Facilities departments regarding issues which may concern technology

- Coordinate as necessary with others on staff or committees on issues regarding technology integration or development

- Create and Manage annual Technology budget

More specific Responsibilities may also include:

- Implement upgrades, expansion, and patches to operating systems

- Provide routine back-up, verification, and restoration of all network servers and workstation data

- Monitor systems, servers, WAN, LAN connections

- Answer all technology assistance requests

- Maintain user accounts and data

- Maintain workstations, network domains, email, and distribution lists

- Set-up and configure Administrative and Academic workstations

- Provide computer systems, video systems, audio systems, and tele-communications systems troubleshooting

- Coordinate with and be available as point of contact for outside vendors

- Maintain warranties and repair information for network, telecom-munications, or other technology-based systems

- Find and manage contractors as backup or aid to the System Administrator

Potential applicants for system administrators should be familiar with current operating environments, wiring systems, and technologies.

If your organization is structured in such a way, a good strategy may be to define a technology committee prior to hiring a system adminis-trator. This committee would have technology savvy individuals as members, and ideally, representation from someone who already manages such types of projects.

Remember that this is not a typical type of systems project. Most environments include multiple systems people: Some in front-line sup-port functions, others in more administrative and managerial roles. This person is a "lone ranger", and will likely have no on-site backup for technical issues, or support. Because there will always be multiple problems and projects going on, time and task prioritization skills are a crucial requirement, as is the ability to manage multiple tasks effectively. Your organization will likely want an independent person who is confident with technology, and confident in making and managing these types of decisions.

Because the system will inevitably move from implementation to maintenance and support, you want to try to get someone who is comfortable with both types of situations. Many systems people are job-hoppers. They like new and different projects, but are not emotionally involved for the long haul. As with any position, your system will run much more smoothly if your system manager had a hand in development of the system. Because of the eventual shift to maintenance, important also is this person's ability to work effectively with the others in your organization. A great deal of this job, especially as time progresses, will be in support and training of this system. This person's ability to work and communicate well will have a long-term impact on efficiency and development.

VOLUNTEERS AS A MEANS FOR SUPPORT

Understandably, not all situations or organizations are equal. You may not be in the market for a system manager, part time or otherwise. For your organization, there are many other good options.

The best option may be to have a volunteer maintaining your system. The key to this is that depending on what is being done, you will still need a screening process and some standards. The best option is to try to find a system consultant who is willing to do volunteer work, or most reasonably, work for materials costs alone.

It is important to realize also, that what is free is not the same as what costs money. Volunteer's time commitment is directly related to other factors in their lives, both professionally and personally. It may be that your system needs to be maintained on the weekends, or other off-times to accommodate the schedule of your volunteer.

With either a contractor, or volunteers, your staff should also be trained in basic troubleshooting skills. Many companies and local community colleges offer basic computer repair, troubleshooting and maintenance classes. Possibly sending a key employee to some training will cut the contractor cost, or in the case of the volunteer, the time that they need to spend doing (relatively) low-level work for your organization.

Volunteers and Project Ownership

One problem that does exist with volunteers, especially in the realm of technology, is a feeling of ownership that comes with setting-up and maintaining a system. Your volunteer may be a hobbyist, or have a certain type of expertise. This can work well for you, as this person is energetic and enthusiastic about your system.

Alas, there is also a downside to this situation. This ceiling of knowledge and interest or time may become an issue as your system grows. There should be a committee of people, or a group internally which is managing specific goals and objectives, just as if this volunteer was under your employ. Starting with this plan early on, possibly prior to engaging a volunteer, may help your organization to keep a better handle on the direction of your system.

Additionally, having some kind of long-term rotation may also help. Volunteers, like paid employees, eventually hit a burn-out period in their path of volunteer service. At some point, after system failures and stress have become a part of this responsibility, this person will not likely have the same level of energy and ideas as they did originally.

If possible, volunteering for this type of job should be for a limited period of time; one year or two years, for example. Understandably, finding qualified people may not be easy, but this may help you to rejuvenate the system periodically, and to get fresh ideas for development.

Technology Committees

Technology committees are the next logical step beyond the single volunteer, and can readily coexist with your volunteer system administrator. They can greatly aid to develop your technology goals and objectives. Technology committees tend to sprout in academic environments, more than in other non-profit organizations. Even if you are not a non-profit, you might establish some kind of committee amongst different levels of people in your organization to discuss technology needs. This can help greatly in the needs analysis work which should be done at the outset of a technology project.

If possible, you want to try to recruit people with expertise from both your community, and a similar community who may be further along in their process. Technology committees, communication between differing groups, even books like this one serve one purpose: To reduce the learning curve, and to help your organization avoid reinventing the wheel.

Especially as your network matures, the technology committee should be a hub of people, rather than an entity unto itself. As a network administrator, you want your policies and practices to be developed internally, not through a group which inherently does not have the same degree of involvement in your particular needs and structure. The technology committee should work with other volunteers and staff to help them identify and meet their needs.

In our organization, we have several different management committees. We have a committee that handles finances and budgeting. This committee has amongst its ranks a great deal of accounting and finance people. We have other committees, such as a long-term planning committee and a technology committee. As many of us are not from the realm of business, and have little experience with budgeting issues, the finance committee has amongst its charge to assign a liaison to each person responsible for a departmental budget. Our departments include the school, the religious education center, and a number of people who head projects for the parish, including technology. The Finance committee members help us to intelligently create and manage our budgets.

In the same way, technology committee sends liaisons to these same people. People in this committee help the staff to come up with ideas regarding our website, and other technology development issues. However, to keep them out of the problem-response circle, *The technology committee should not be expected do nuts-and-bolts repair of equipment.*

Nonetheless, Our Technology Committee has been invaluable. Something to realize about technology committees, is that their function and value changes over time. Originally, our technology committee was designed to figure out what the needs of the organization were, and to draft a very high-level plan about services and features which should be incorporated into a Technology System.

The membership met with various departments, and drafted a document, which would help to guide us for the next few years. This document was a resource of ideas, and expectations of what this system should do.

Fortunately, over time all elements of this plan were realized. At this point, the technology committee was forced to reinvent itself. We even considered completing its mission. After discussing options with our staff, we decided that even in a maintenance and review mode, that there was still a clear role for the committee to play. Our organization is currently in a development and upgrade phase, and the committee is gearing our users up to utilize the technology.

Our committee has grown to be a hub for our website, which is an area that sorely needs interest and development. This is not to say that this task will be the last of the technology committee, or that this is a job to be "done". The committee will work on developing this aspect of the system for a while, and will remain coordinated and active until other projects come up. One good idea to keep initiative levels as high as possible, and to keep people interested and involved is to create a list of goals and objectives for a coming year. Reproduced below are our objectives and mission statement for this year:

Goal: *To develop/implement/maintain an attractive web site that is a source of up-to-date information about St. Joseph Roman Catholic Parish, to include all groups, programs, education, and ministries.*

Objectives:

1. To give a creative, uniform look to the "index" web pages of St. Joseph's groups, programs, education, and ministries.

2. To set up procedures, guidelines, and templates that allow the content of the various pages to be updated regularly by parish staff, groups, programs, education, and ministries representatives, while the overall "style" of the site is maintained by the web master.

3. To provide training opportunities for parish staff, groups, programs, education, and ministries representatives regarding web page design so that they can develop and maintain the content of their respective pages, within the procedures and guidelines set by the web site subcommittee.

4. To provide input into the annual parish budgeting process with regards to ongoing expenditures to maintain and update the web site.

As you can see, our committee's goals and objectives amount to a one-page document. This document combines a higher-level mission statement as well as fairly specific goal statements relating to its achievement.

CONTRACTED SUPPORT

In tandem with volunteer support, or even the hired system administrator, Eventually your organization will want to do something that may require contractors. Contractors allow fresh perspective, and the kind of specific expertise that you cannot expect of your internal people. Contractors are also less likely to be mired in, or influenced by, internal politics or practices. Contractors can make a lot of technology projects work in situations where they would otherwise fail.

As Contractors are expensive, they may not be a good answer for comprehensive support in your organization. They do perform certain specialized functions to a higher degree of quality than your own people may be able to accomplish for the same cost.

For instance, one of the first initiatives of my administration of our network was to implement a high-speed Internet connection. Our organization decided that Internet access and group email was important enough to warrant a high-speed DSL connection. One side effect of an "always connected" strategy is an almost constant barrage of outside attempts to attack our network gateway, and our machines connected to the internal network.

This can be a very dangerous situation, compromising both the satisfactory operation of the system and the protection of your users' data. I realized that we needed to implement a firewall. A Firewall is a piece of equipment which is placed in-line between the interface equipment belonging by the Internet Service Provider, and our network's own equipment. It's sole purpose is to act as a protective barrier between the outside world and the inside network.

I had no prior experience with firewalls, or with firewall configuration, and did not feel adequately competent in this area to make the technical decisions without some outside council. I contacted several vendors and network integration companies, and learned a great deal about network security, an industry unto itself. Network security and web presence, for many networks, are two projects which require outside assistance. There are many issues surrounding obtaining and installing the communications equipment, the network protective equipment, and configuring a working web server.

My advice is to look to contractors as keys to open specific doors. This is mainly because of cost. For instance, our network security company, bills for its services, at the time of writing, around $250 per hour. A typical session with the contractor costs our organization between $750 and $1000. Considering that the firewall equipment itself was, when new, around $5000 to purchase, the post-sale service will soon exceed the cost of the equipment.

Bear in mind that while many aspects of data-processing equipment decrease in cost over time, specialized equipment and skills, by nature, will command a premium. Network security, right now, is the top priority to many network administrators. Possibly, next to backup, it is the most important priority. While most contracted work will not cost as much per hour as network security, the network contractor is like any skilled tradesman, like the plumber or the electrician. You want to hire someone who is available when you need their services, and get them out the door as quickly as reasonably possible.

Contractors work differently, depending on size and the scope of your project. Most contractors of any size are set up to accommodate you in two ways. The first is called "Time and Materials (T & M)" costing. Time and Materials costing can work well for small repair projects. This is basically a charge, per quarter hour, for the amount of time that the contractor is on site. It is likely that you will also be charged a reduced or similar rate for travel time of the contractor, especially if there is a lengthy travel time involved in coming to your site. In addition, you will be billed for the amount of all of the materials used to make the repair. Usually, the contractor will arrange all of the parts used in the repair. Although less convenient, some contractors will allow you to provide parts, if this is required by your organization, a move which can save your organization money. If you are bringing contractors in to repair equipment, time and materials is a fair and reasonable way to arrange contracted service.

The second means of billing is called the Retainer. A retainer is a common practice in the realm of professional services. You are billed for a certain amount of "time", usually at a reduced rate derived from the firm's regular hourly service rate. The rate should be reduced somewhere in the neighborhood of 10 to 15%.

Retainer means that you must purchase a minimum amount of time, sometimes with a provision that it be used in a certain amount of time. You will want to get this information up front. Depending on the type of work involved, retainers can be, in the long run, a better value for your organization. Retainers typically start, as of this writing, between $1000 and $10,000 depending on how many hours you purchase, and how quickly you want response after a problem has been called into their response center. Many companies will have a sheet of pricing options. Many will also have other premium options, like 24 hour/7-day week/365-day per year support. For additional money, they can host certain aspects of your network, or provide technical people on your site for a certain amount of time per week or month, to act as system managers. If your organization is so inclined, they can use a retainer to basically "hire" a system manager without having to actually hire an additional employee. Unlike many types of service contract, most retainer agreements with technology companies are arranged so that after you run out of time, there is no further obligation.

ABC Computer and Telecommunications Contractors, inc.
PRICE SHEET AS OF JANUARY 2002

Regular Hourly Pricing for on-site technical support..................................$100/hour

PLAN	PRICE	HOURS	RESPONSE GUARANTEE
"Bronze 24 Plan"	$1125	15 @ $75.00/hr	24 hour guaranteed response time
"Bronze 3 Plan"	$1725	15	3 hour guaranteed response time
"Silver 24 Plan"	$2100	30 @$70.00/hr	24 hour guaranteed response time
"Silver 3 Plan"	$2700	30 @$70.00/hr	3 hour guaranteed response time
"Gold 24 Plan"	$6500	100 @65.00/hr	24 hour guaranteed response time
"Gold 3 Plan"	$7100	100 @65.00/hr	3 hour guaranteed response time

FIGURE 3.1: ABC PRICE SHEET EXAMPLE

Contracts, Retainers and Time and Materials Costing Arrangements

New projects should usually be arranged using a contract, rather than a retainer or T&M type billing. New project contracts, unlike retainer or T&M, allow you to predict and basically guarantee a fixed price for labor, and materials. This is very different from work done using a variable time scheme, like a retainer or T&M, where every hour of work costs your organization money.

Be wary of companies who try to execute large projects using either retainer or T&M billing. Companies who want to do this may be inexperienced in undertaking large installation projects. This can cause your organization to experience problems related to design or installation quality. Alternatively, a company who tries to do a large project using time and materials billing may be trying to cheat your organization, using a relatively costly billing scheme.

Service Contracts

Service contracts take several forms. You may have purchased expensive equipment, or equipment who's operation is critical to the continuation of business in your organization, like a network router, hub, or server. This type of equipment should always have a service contract, preferably one that is "no fault", if available. Especially for key equipment, service agreements should be arranged that specify a 24-hour replacement policy.

Equipment that should have Service contracts:

INFRASTRUCTURAL EQUIPMENT, SUCH AS

- Switches

- Hubs

- Routers

- Shared Printers

- Servers

- Backup and tape drive equipment

CLIENT EQUIPMENT, SUCH AS

- Portable Computer Equipment

- Specialized equipment, such as envelope printers or scanners

- Digital cameras

- Audiovisual equipment, like video projection equipment

One of the first groupings of equipment that I had placed on service contracts was my shared printer installation. My network has 14 shared laser printers, all with on-site service agreements. Laser printers are the most mechanical of network devices, and given the right amount of use, will eventually fail. Our service contract includes next day on-site repair, and regular scheduled maintenance, which is also important.

A different kind of service contract is a type that covers some type of continuing service rather than a correctly working product.

Two types of arrangements, which fit this description, are hosting of your web site and the leasing of email service to your organization. There are many companies who will run your email and web site for a fixed amount of money per month. This amount of money varies depending on the number of email accounts you require, and how involved of a web site you need. For additional fees, some firms can also handle all of the development activities that your organization requires. It may be necessary only for you to give the "content" (The actual typed or graphical information) to the web hosting company (the company that runs your site) in order to maintain a professional web site.

Your users benefit from the services, while moving the burden of responsibility over the mechanics of the site to the web hosting company, a group of people for whom running web sites and email is the core competency. The hosting company deals with issues like network reliability and security, and they have the high-speed connections required to provide reasonable speedy access to your site from the outside world. All that must be maintained on your end is the Internet connectivity. In addition, the cost of hosting is usually lower in the short term to maintaining the equipment yourself. If done with some foresight and planning, there is virtually no downside to this arrangement for the small organization.

PROACTIVE AND REACTIVE SUPPORT OPTIONS

Many instances of problem response in an Information Technology installation is initiated by the user of the system. For instance, Louise, the secretary, will come into work on Monday, and will not be able to print. This accounts for much of the laborious work of a network administrator. Through some failure in some part of the system, Mary is unable to print a document. She calls the network support person, who comes to her desk, and fixes the problem. OK, Problem solved.

This is called *reactive* response. Reactive response is not the ideal situation for the user. First and foremost, the user sees a visible breakdown in service. Whether this is the fault of the network manager, the hardware along the line, or the software is of no interest to the user. The psychological impact of *reactive* response is the same as when your car won't start in the morning. Ninety Eight days in a row you start your car, and you go to work. You think about the news, or listen to music, or prepare yourself for the coming day. You are not thinking about gaskets, pistons, or gasoline. On the ninety-ninth day, your car won't start. "Maybe it's time to get a new car", you think. You plead with it, play with it, and it eventually starts. Ultimately, though, the trust has been broken. It may take a trip to the repair shop, or a week of successful starts to regain this trust.

Your network is the same way. People using your computers think in terms of service, not structure. They think about sending mail, talking on the phone, getting work done. They do not appreciate that even well-run systems are down 1% to 5% of the time.

The more times that the network won't respond as expected, the more times that the trust is broken, which results in unhappy users. Unfortunately, they will begin to associate this unhappiness with you, the visible person who is responsible for the network. It's just the same as payroll. The first week that payroll doesn't send out a paycheck is the last week of peace between employees and accounting for a while.

REACTIVE TROUBLE RESPONSE SYSTEMS

Problem response is a way of life. You need to have a number or process in place to resolve problems. Computers fail, and nothing short of perpetual service can prevent this from happening.

No matter how competent or complete your problem resolution system, you need a connector between the problem resolution people and the clients. Whether you have all volunteers, a system administrator, contractors, or a mix of these types, you need the bridge the gap between the people having problems and the people who fix the problems.

Problem response is more complicated than it may seem on the surface. As your organization becomes more involved with technology, the need for accurate, effective service response will increase. Organization and recording of service requests are the keys to successful client service.

My recommendation is to try to work with plant and facilities groups to build one system. This is what I initiated here. Technology problems are not unlike other facilities related issues, and one system for problem response is better than two.

I designed our system here, and I think that it works pretty well. Our system is a mix of paper forms, an electronic system, and policies regarding time frames and response times.

The plant and facilities people and myself meet weekly on Monday mornings to discuss things having to do with the week's activities, to check each other, and to talk about "open tickets". Open tickets are problems which have not been resolved. In our organization, there is no managerial connection between facilities, business management, and the technology department, but we try to work together as a team to get things accomplished.

Our work order system works like this: The employee fills out a request form from their stack of blank forms. The completed form is given to a secretary who manages the work order system. The secretary enters all of the information into a database which is located on the network. Each of us take a part of these tickets (the Operations manager deals with Heating and Cooling; the Plant manager deals with facilities repairs, and I deal with computing, Alarm and A/V issues). The database routes the ticket automatically into a list for each of us, and we can, at will, print out a list of either all the unresolved tickets in the system, or just a list for our own unresolved tickets.

Upon resolution of a problem, we use the database to print out the form (with information included), sign it, and return it through inter-office mail to the person who requested the fix. This completes the circle, and lets the person know that their problem has been addressed.

While this type of form may be overkill for your organization, we chose to do it this way because it creates a paper trail for problem resolution. The form is simple enough that there is no real problem completing the information. Most of the choices are printed on the form, and need be circled to indicate location and topic of the work order.

There is also a place for comments, which are inserted into the completed ticket from the database. These comments can help the requestor to understand why the problem occurred in the first place, or why a problem could not be resolved.

The prime purpose of these "reactive response" systems is not for day-to-day maintenance. If you are the sole system administrator in your organization, you will quickly realize that there are always more spoons than you have hands to stir. When a real problem does arise, such as a network outage, you do not want people calling your office because they can't print or because they have a relatively small problem. The trouble-response system should work to give you a buffer to do your work, and also have the clients feel as though they have a system by which they can resolve their problems.

Trouble-response systems should have a policy behind them, one that specifies a time frame for normal resolution of problems and explains how the system works.

The system that we implemented at St. Joseph's involves a database back end. This database allows the individual "trouble tickets" to be tracked, and to be analyzed or distributed. As many people use this system, to input information, and to receive trouble tickets, a multi-user network friendly database was required.

<u>Analysis can include:</u>

- How long was a ticket open

- Was the ticket actually resolved, or denied for some reason

- How many tickets does each person have open, currently

- How many tickets has a person entered this year, over the entire scope of the system

- How many tickets have been opened for a specific problem or type of problem

A Telephone-Based Helpdesk Model

Your organization may not have a need for such an involved system as the one above. A very simple way to manage requests for help is to set up a phone line, or a voice mail box, where the people can contact you. The mailbox might specify in the message what type of information is necessary, such as a person's name or a description of the problem.

The mailbox is inherently less coordinated than the work order system based in paper or electronic format. The convenience of the mailbox for the users is that it requires no administrative overhead. People leave you the messages, you pick them up and follow-up with the problems. You may choose to keep a pad of paper, or you own database to keep track of problems that have not as yet been corrected.

Mailboxes or answering machines are, additionally, less of a barrier for use. Asking people to make a phone call is a simple request. If this is an automated system, the voicemail should be consistent, and always answer in lieu of an actual person. Voicemail boxes can also be configured to allow a message to be changed, so that if a service is not working, people will become informed of this when they call to report the problem.

PROACTIVE MAINTENANCE AND EQUIPMENT REPLACEMENT

The other side of system support is the proactive type. Proactive support is really the only way to alleviate service problems in your organization. The first priority of any support function in an organization should be to minimize the amount of reactive support, rather than to improve the reactive support function. While realizing that reactive support is always going to be a theoretical reality, organizations should seek to minimize its activity to whatever degree possible.

The only way to fight problems in your technology system is through preemptive maintenance. Preemptive maintenance may be the regularly scheduled cleaning of office equipment, the scheduled replacement of *mechanical* parts, such as power supplies, hard disk drives, and replaceable clock batteries on client equipment.

Power supplies, and especially hard disks, are the two weak points of a computer system. Their nature is to eventually fail, due to continuous use when a computer is functioning.

While for most organizations, the idea of replacing working components would seem ridiculous, any hard disk drives and power supplies whose age exceeds more than five years should be replaced in most circumstances.

Hard disk drives, especially, have fixed life spans, as stated by the manufacturer of the drive. These are expressed in a number of operating hours, called MTBF, or "Mean Time Between Failure" rating. Newer drives, such as those that are less than five years of age, are significantly more reliable, and as such are rated much higher in terms of MTBF, some with ratings such as 500,000 hours between failure.

MTBF is designed to give you an approximate idea of the life span of a hard disk drive, so that you can get the old drive out through controlled means prior to its failure. Some drive failures may be inconsequential to you. Others, such as server hard disks, may be of greater concern.

Cleaning, and at the same time, the regular replacement of batteries, should be performed at least annually. Our installation is located next to an active limestone quarry, and requires semi-annual cleaning because of the higher levels of particulate matter in the air. Depending on environmental conditions in your immediate area, you may need to adjust this number accordingly.

Lithium clock battery failure in computers accounts for a great deal of glitches and unexpected problems, and strange error messages for users to decipher. Changing these batteries will help to reduce these types of service calls, and will ensure that you have batteries on hand to do the replacement.

The best advice is to keep an eye on problems that seem to crop up regularly in your organization, and to add those items necessary to your list of scheduled maintenance.

Regular Maintenance

Most networks require a certain amount of baby-sitting. Doing these items regularly can prevent a great many problems. The side effect of doing regular maintenance is that you will have a finger on the pulse of your system. You will notice the beginnings of a lot of problems, and can resolve them before they become noticeable to users.

Regular Maintenance Items

• **Daily Backup Tape rotation**

See the section about regular data backup, but this point cannot be overstressed.

• **Daily Monitoring of the event logs on the Servers**

Most computers have some kind of event logging system. There is usually some kind of viewer whose purpose it is to display the events in the log. The event log will show all of the activities that have occurred on the machine. You are not looking for anything in particular, just to get a picture of what has happened recently. Is there an application error? Is someone printing something ten times in a row? Server problems usually manifest themselves by showing error messages in the event viewer. A lot of messages in the event viewer are normal, and even many of the more serious warnings are normal for a server to have. You do want to make a note of any "serious" messages from the a server.

Daily Checking the event log on the Mail Server

The event log for a server used to send and receive internet email is particularly important to check regularly. Look for any connection error messages or send errors. These errors will indicate that some type of problem with the transfer of mail. Checking this server is particularly important because due to the nature of the multitude of links in the mail server chain means that problems can occur that do not get detected for days.

Daily Checking the event log on the Tape Backup Server

Checking the event log on your tape backup server is a simple, but important, activity. Tape backup programs usually display error messages and results in their own program.

Look for "FAILURE" status readings. This indicates that the previous day's backup did not run successfully. This, can mean a problem with that day's backup tape or with the backup definition instructions. Note that a failure may mean that a shared folder is no longer accessible to the backup server.

- **Weekly Cleaning of catch-all email accounts and log files**

The administrator account on the email server is usually used for a few purposes. Log files, denied mail, and "lost" email files show up here. It fills up rapidly, and needs to be emptied on a regular basis.

Failure to empty this catch-all account can result in the server eventually choking on its trash.

- **Weekly cleaning of Tape Backup Drives**

Tape drives are expensive and finicky devices. Cleaning the drives regularly is important to ensuring a long life.

- **Weekly Monitoring of the Anti-Virus statistics**

Keep an eye on the antivirus updates, which sometimes will not run. It is important to maintain up-to-date copies of all your antivirus system files.

Additionally, there are other items which may need to be addressed on your network, and should be scheduled on at least on a bi-annual basis:

These include:

- **Replacement of consumable parts (rollers) and cleaning of printers**

Changing of Rollers and cleaning brushes in laser printers. Cleaning of the platens in the inkjet printers and impact (dot-matrix) printers. Printers are mechanical devices, keeping them clean ensures clean copy, and a long life.

- **Changing of toner cartridges**

- **Physical cleaning of servers, fans, and fan filters**

- **Physical cleaning and demonstration of functionality for client computers**

  - Is the mouse and keyboard working well? Mice and keyboards get dirty, and eventually need to be replaced. They are almost classifiable as "consumables".

- Is anything obviously broken on the computer? Are all switches, keyboard keys, etc. present and working?
- Are there any odd or unexplainable error messages?
- Is the computer able to print to the right printer?
- Is the Sound working properly?
- Is the computer connecting to the network?
- Is the time correct? (Incorrect time can indicate a dead battery).
- Does the desktop look OK? Are the icons arranged?
- Does the display look OK? Is the Color good?
- Is the machine making any strange noises from the drives or fans?

Any device with a fan, whether it be a printer, computer, or other network device needs to be cleaned regularly. Regular cleaning prevents dust build-up that can cause shorts, static electricity, and other ills to sensitive electronic equipment.

PARTS SUPPLY

Some surprisingly large organizations work reactively in the area of supplying parts. Rather than ordering a supply of parts that fail regularly in any system, these organizations wait for a failure to occur, then are out of service for days waiting for the parts to arrive.

While it may not seem to be a good idea to sit on an inventory of parts, all of which lose value over time, it is necessary to realize the difference between support goals and sales goals.

In a sales-based organization, like a computer parts store, it is advantageous to stock as little inventory as is reasonable and possible, operating rather on the basis of ordering less, but more often.

To some extent this rationale makes sense. Most parts in a computer system are basically commodity items, and decrease in cost as a factor of value over time. The $80 CD-ROM drive last year will cost only $65 if purchased today. Next year at this time it may cost only $55.

The balance in an organization which does not sell parts or service, such as a non-profit, is to make sure that there is enough surplus on hand to fix equipment. Most non-profit organizations for which I have worked, do not expire a great deal of their equipment. Most of the equipment is kept on hand, for later use. As time goes on, parts to repair these machines may be less available, and thus may take more time to get, or be more expensive.

Maintenance Parts:

- Keyboards
- Mice
- Monitors
- Power Supplies
- Fans
- Hard Disk Drives
- CD-ROM Drives
- Floppy Disk Drives
- RAM memory
- Power Cords

Consumables:

- Printer Paper
- Printer ink and toner cartridges
- CD-R's/CD-RW's
- Backup Tapes
- Floppy Disks

Having roughly 10% of each of these sets of parts on hand based on the total number of machines will ensure that no ordinary problems will fail to be resolved in a timely manner.

TRAINING

The best form of preventative response for support is through the use of thorough training. Training is a key aspect of system management, because without it there will be very little productive use of the services. You will find that at every level, from your clerical staff to the executive level, users need to be trained in the proper use of the system programs.

A good first step is to have a training class covering the use of the operating system. Provided that you are using a standard type of operating system on the client computers, this is fairly easy to do. As a computer savvy person, you can arrange a "basic training" class, which will probably cover very little new ground. This is a good first step, not because it teaches a great deal of new information, but because it gives you a feel for being the teacher. Additionally, it gives others in your organization the feeling of being students, which may not be an easy role for everyone. Because you are covering basic territory, most of your staff will become comfortable with the idea of training not as a negative "classroom" experience, but as an opportunity to learn new things and be a part of a group. In addition, you know that everyone is now at that "basic" level of knowledge, which is a good starting point for your next level of training.

Soon after your first training class, you should begin training in other basic-level applications. This includes teaching standard applications, like word processors, spreadsheets, and database products. These products are likely to be widely used in your organization. While slightly less commonplace than the operating system use, these types of programs are still probably familiar territories for most people.

Following this, you may wish to either go into more advanced levels of training concerning particular applications, or you may need to go into training for custom applications.

If you are uncomfortable with training, there are many good companies out there that teach these types of programs. Whether you teach yourself, or have the training done by a third-party educator, you want to try to structure your training in a familiar, consistent fashion. If using a third party, it is a good idea to try to get the same instructor for subsequent training classes, rather than a "revolving door" of training providers, or instructors.

As your staff begins to become familiar and comfortable with the instructor, subsequent training sessions will be more open and interactive. Your staff will find it easier to voice questions or concerns about the material, and you will find that the training is wholly more effective.

Becoming more commonplace is the idea of instructor review. Most reputable training companies, and colleges and universities as well, do a follow-up review with the students after the training session has been completed. If you do the training yourself, you should some form of follow-up as well. People may not be as open and honest with you, if you are a volunteer or staff member yourself, as they would be with a contractor. Giving everyone an opportunity to voice their opinions about the quality or effectiveness of a training session can help to improve future sessions.

Basic Training

The key to managing training is to identify the core set of skills which everyone in your organization needs to have. Like many other things in technology, these skills are like wheels, they need no reinvention. Skills can be listed like basic tools in your toolbox. These issues are mentioned Below.

Computer Hardware Familiarity

Your users should have a basic familiarity of how a computer works. This is a good starting point, because you can show them the parts of the computer, the peripheral equipment. Mouse and keyboarding (touch-typing) may be addressed generically. Although mouse skills are important, touch-typing skills are best left to education centers which are set up for this purpose. It is important that people who do not know how to touch type are taught correctly, as it is much more difficult to re-teach this skill, than to teach it correctly from the beginning. Touch typing is a process not learned in a two hour session, it can take weeks to learn to begin to type.

Elements of The CPU

> Using Floppy Disk Drives
>
> Using CD-ROM drives
>
> Drive Lettering Practices (in an IBM environment)
>
> Starting the PC
>
> Shutting down the computer
>
> Explanation of RAM
>
> Explanation of Hard Disk drives
>
> Explanation of the various fans in a computer

Elements of The Display

> Powering up the display
>
> Changing hardware settings on a display
>
> Healthy Environment:
>
> Display problems, such as flicker, which are caused by outside sources, (eg. Close-range Fluorescent Lighting)

Printers

    Ink-jet printers

    Laser Printers

    Connections to computers/LAN

    Possibly basic troubleshooting, like changing cartridges or clearing paper-jams

Networking

    Network cards and wire

    Basic explanation of how the network works

Using The Operating System

Teaching a class on using the operating system? Many people's sole use of the operating system is as a springboard to open other programs. The operating system is superficially familiar to most people. This being said, there are many aspects of controlling the computer which elude many savvy computer users. Teaching basic skills can greatly reduce the number of configuration-type trouble calls that you receive in the future.

The majority of operating systems do several things for the user. They offer File Management, Program Management, and the ability to configure certain variables of the computer's configuration. In a managed environment, it may be unnecessary or undesirable to teach users about configuration changes such as adding hardware to a computer. Do not assume, though, that because you have not taught these skills, that your users are not familiar with these practices. Most of us have PCs at home, and many of us love to tinker with them.

Using the File Management Tools

    Navigating the file system

Making a folder

Renaming a folder

Making a new text file, Word processor file

Formatting a diskette

Using Printers' Queues

Using CDs and Discs

Formatting a Diskette

Programmatic Options

Program Selection

Program Installation

Launching Applications

Control Panels

Changing Sound Options and Volume

Changing Color Schemes

Setting a New Background Pattern or Screen Saver

Using The Computer System's Find Feature

Re-setting the Time, Date, and Time-Zone

Network Features

Resetting Passwords

Logging into the Network

Connecting to Network Drives and Printers

Word Processing

Word processing is something which most everyone who has used a computer has experienced. However, although it is a good example of a common ground application, most people use very few features which most modern word processing software has to offer.

Below is a sample of what you might offer in a word processing class:

Working with Documents

> Creating a new document
>
> Finding and Opening an existing document
>
> Saving a document
>
> Using Templates
>
> Print a document and using Print Preview

Text

> Use the Spelling Checker
>
> Grammar Checking and Thesaurus
>
> Highlighting Text
>
> Cut, Copy, and Paste
>
> Using Typeface Styling (Bold, Italic, Outline, Underline)
>
> Select and change font and font size
>
> Find and replace words or sentences
>
> Insert date and time
>
> Using Bullets

Working with Paragraphs

> Using Justification (Center, Left, Right and Justified)

Setting line Spacing

Using Borders

Use Tab-stop settings

Set tabs with leaders

Set margins

Create and modify headers and footers

Creating Columns

Using Tables and Graphics Within Documents

Create and format tables

Add borders and shading to tables

Inserting Pictures

Using Clipart

Spreadsheets and Databases

Spreadsheets are a lot like Word Processors in that most people who have used a computer have used one from one time to another. There are basic features which should be reviewed, and explained. Chiefly, education should be stressed on the proper use of the spreadsheet. Spreadsheets and Databases, although similar in appearance, are not the same. Using spreadsheets for purposes where a database application is needed can cause a great deal of inconvenience.

Email Software

Email is a very basic service, one that has become all but ubiquitous. Very likely, you are using some version of one or two mail programs. Mail programs do not vary greatly, and most standard topics can be taught:

Many good email products also include other features, such as journal functions, and calendaring programs. As feature sets vary from program to program, this outline does not address these particular topics. This should not detract from their presence or value, as they can be usable tools in your system's arsenal.

A Sample Email Course Outline:

Email General Topics

   Email addresses and domain names

   Basic information about how email works

   Connecting and Logging Into Your Group Email System

Sending Mail

   Making a New Message

   Attaching Documents and Pictures to Email Messages

   Message Size Considerations

Receiving Email

   Using the email program to download and view new messages

   Opening Email messages

   Downloading Attachments

Email File Management System

   Creating new Folders

   Moving Email into Folders

   Deleting Email

   Creating rules to Automatically file your mail messages

Addressing

Using The Address Book functionality

Using Distribution List

Using Global Distribution Lists (These are central lists which the system manager creates on the server for everyone's use)

On-Site and Off-Site Training Options

Off-site training can be a more effective option than equivalent on-site training, especially if you are conducting intensive education or if the product is inherently complex. Dedicated classrooms may be set up in a fashion that is more conducive to learning than what can be reasonably arranged at your facilities. Typically, on-site training is conducted in libraries, classrooms, or meeting rooms. Meeting rooms in your organization may be difficult to reserve during the times when training is available. The acoustics and lighting of many meeting rooms is not ideal for training. Chiefly, though, the problem is the room's location: The meeting room is on-site. On-site training has a real setback, it allows for intrusion and distraction. So-called "emergencies" are more likely to drag a person out of a training session in a meeting room, much more so than if the person is off of the premise.

On-site training can certainly be successful. Many people are more comfortable with an on-site training arrangement, since they will not have to drive anywhere out of the way.

Users should never be trained in the office. This type of training is a certain recipe for failure. Distractions are frequent and numerous. In the office, telephones ring, and other stress is close at hand. While training, other responsibilities need to be put aside. Although of great benefit, training is an intrusion. Managers may view training as a part of the day or week for which no production is evident. In the minds of many people, you have taken some of the productive time away, or forced them to work late to complete something on schedule. Be wary of this attitude, it is unfounded. Training is a valuable and worthwhile activity in any organization, non-profit or otherwise. Although your staff is busy, time needs to be taken to achieve a literacy in technology.

Maintaining program consistency, as well as choosing a vendor with a high degree of quality and thoroughness is key to your education initiative. Providing training sessions that are perceived as being poorly executed will surely reduce buy-in for subsequent sessions.

Education and Meeting Space

Some organizations will undertake more extensive training programs than others. Depending on your level of technology commitment, and funding concerns, you may choose to do funding either in-house, or at the educator's off-site location.

Multimedia meeting rooms serve a dual purpose. Meeting rooms outfitted with technology equipment can bridge the gap between the computer as an information tool, and your traditional means of presenting information. These spaces can also be used as traditional meeting spaces, or for education efforts. Having a room which can be used for training expands your options to those educators who have no facilities of their own. As with many aspects of technology management in smaller organizations, use of independent contractors can reduce costs of training, which is an expensive undertaking. If possible, educational and meeting space should be planned out before any building projects are undertaken. Meetings and seminars are usually integral to the work of any non-profit organization, and their design is very important.

Meeting rooms in an organization should be furnished comfortably, and should be able to be subdued in terms of lighting and acoustics. For the purpose of technology use, it is important to be able to darken the room, and to have all of the people face a projector screen on the wall.

Factors to consider when planning a meeting space:

• Comfortable Furnishings

- Comfortable lighting, with at least three degrees of light for normal activities, a subdued level for presentation activities, and a dark level for use with projectors.

- Adequate room for the number of people expected without feeling "cramped"

- Adequate ventillation

The key to a multimedia room is the data projector. Data projectors have evolved over the last decade as possibly the most versitile pieces of equipment in an organization. Data projectors allow Computers, VCRs, Cable and TV Tuners, videodisc and DVD players, and other forms of audiovisual equipment to be projected onto a screen.

With the use of wall-mounted speakers, appealing sound can also be included in the arrangement.

The video projector is connected to a switchbox which is located on the podium. A computer, VCR, and other A/V equipment is installed into the podium as well. Using the switch box, mouse, and keyboard, a person can conduct a multimedia presentation using the tools in the meeting room.

Obviously, your meeting room may not need all of this audiovisual equipment. A good arrangement may be a VCR, a wireless microphone system, and a computer installed so that all of the components can be heard through the speakers in the room.

Usually, the projector is installed into the ceiling, so that color balance, image size, and focus are pre-set for use with a minimum of tweaking when the room needs to be used.

WINTER: RENEWAL AND REBIRTH

○ ○

"The more you know, the less you need"

—*Ancient Aboriginal Saying*

The window air conditioner whines and hums in the background. Fleetingly, I wonder if the air conditioning will last the rest of the summer. I also wonder about the new tables, if they will arrive in time, because I haven't heard from the cabinetmaker in several weeks. The thought passes quickly as I look at the drawings again, calculating the amount of metal piping [Metal Clad Conduit] will be needed. Strewn across the floor are boxes of connectors, no less than three power drills (two are Ed's, one mine), a saw, and at least 25 10' Grey hollow metal poles. Ed is the contractor, a good natured man of about 60 who peers over reading glasses as he examines and notes the imperfections in the wall. A master brick-layer by trade, he stands at the wall with a marking pencil and a wooden folding bricklayers' ruler. He taps the wall judiciously with the ruler, and makes a careful "X" where one of the data jacks should be. He embodies the craftsman's adage "Measure twice and cut once". Ed and I have decided to use metal electrical conduit and matching metal boxes ('nineteen hundred boxes', he tells me they are called in the trade), for the data wire, to match the ones already installed for the electrical service. Ed and I have been talking all morning about what kind of screws and shields will be necessary to hold the conduit to the cinderblock wall. I never realized that there were so many choices involved with building materials. I am in the school of life experience, and this should be an interesting year.

Is your job done? The short answer is "no". The job of the technology manager is never really done.

Early in the book I said that raising a network is a lot like raising a growing child. Your system will grow and, like most of us, become more seasoned with time. Like children, with or without discipline, they'll act up and get messy from time to time.

Your system will require a lot of attention and maintenance through all four seasons of its life cycle. As new network or building additions are made, the system will need to change with the times. It will enjoy a long, productive life will make your organization sing with productivity and efficiency. Some of your people will even begin to like using it, I promise.

Eventually, your organization will need to upgrade its systems. This may be long in the future, considering that your data system may be shiny and new and *fast* as you read this. Make no mistake, the day will come when the equipment is dingy, noisy, slow and less reliable than you remember.

THE VALUE OF INFORMATION

If there is one message that you take from this book, it is that information and not the physical systems is what is truly important. Your information is exponentially more valuable than your infrastructure.

The first facet of counteracting obsolescence is to make certain that your data is standardized, to whatever degree possible. Granted, most all databases are not going to be able to be standardized. Databases tend to work with data in a proprietary way. Unfortunately, when choosing a database package, the standard by which it operates is usually not a high-priority factor in the decision. Staying with cross-platform standards, like SQL is a good idea whenever possible.

On the other hand, Word processor documents and spreadsheets are the easiest to convert and transfer, as are files that are based on *vendor neutral* standards. Vendor neutral means that the file standard is used by many different manufacturers or is in the public domain, so it is impervious to the whims and changes of a unilateral company or software developer. To whatever degree possible, file format standards should be used, to protect data from becoming victim to obsolescence.

Using standard document formats is good for two reasons. First, it allows many people, especially those not using the same exact software, to be able to retrieve data. If you have multiple kinds of computers on your system, this can be very important.

Secondly, and possibly more important, using a standard document format prevents the aging of the computer to prevent the data from being able to be transferred. If you are using a computer system that is several generations behind the current state-of-the-art, you may find it difficult to translate information directly from the old computer to the new. If the data is of enough importance and needs to be converted, you may find success by using the services of a professional data transfer company.

To be a standard, a file format should not be directly associated with one particular program. For instance, Microsoft Word may be a standard amongst Windows word processors, but it should not be considered a "standard" format, because it is backed by one company and program only.

Examples of standards:

General Standards for Text and Graphics Documents
Adobe® Acrobat™
Adobe® Postscript™

<u>Word Processor Standards</u>
ASCII Text
RTF

<u>Graphics Standards</u>
JPEG graphics
GIF graphics
Tagged Image File Format
PICT (Macintosh Standard)
Encapsulated Postscript (good for high-quality graphics)

<u>Movies and Sound Standards</u>
MPEG for movies
MP3 for audio files

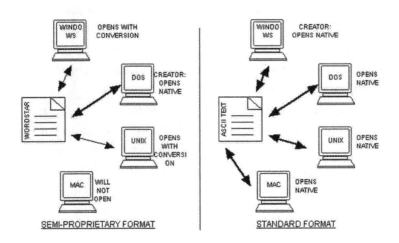

FIGURE 4.1: THE VALUE OF USING STANDARD FORMAT

Obsolescence Management

Obsolescence is a word full of negative connotation, one that I have never liked to use. I have a computer that is a decade old. It still does the same things it did in 1992, and still just as well. Because of its wonderful page-length display, I used it to draft a good deal of the book that you hold in your hands. Sure, Internet access is a bear, and it isn't as blazing-fast as the new machines out there, but I like it just the same, and I'll continue to use it until there is some good reason to stop.

Many people look at their computers as they do their cars. Either there is a never-ending string of expensive upgrades on the path to some "Rube Goldberg" nirvana, or else they are on the perpetual three-year lease program. I think that both of these directions are dangerously shortsighted.

Upgrading equipment in some quest to remain competitive is, ultimately, a losing battle, and ultimately a waste of money. The evidence is clear. In the 1980s, most PCs were of the IBM AT-type computer. The AT computer was a large rectangular desktop cabinet with at least 3 spaces for disk drives, usually more. AT computers also had huge power supplies that could accommodate large amounts of add-on equipment. The AT standard was first released in 1984, and thrived until several years ago. In a world of USB and notebook computers, standards such as floppy disks, serial and parallel connections, and the AT cabinet have all but faded into oblivion.

Computers were sold much differently than they are today. They were built with usually, 5 or 6 of their eight slots available for expansion. It was unheard of, even in early laptop computers, to sell a system which did not have provision for expansion cards. This gave the user the opportunity to upgrade RAM, add a sound card, upgrade their monitors or video graphics cards. They could even add increasingly efficient modems to gain connectivity.

Over time, this arrangement has changed. Computers have become generally significantly less costly. PCs and hardware that used to cost thousands of dollars now costs hundreds of dollars. Computer equipment is no longer the expensive mystery that it once was, and many more people and companies are involved in the industry. This economy of scale has had a lasting effect on technology. Computers are more like VCRs and Toaster Ovens, and the typical user has not the desire to tinker with the equipment. Many larger support structures use third party service providers for technical support, rather than fixing equipment in-house. In-house repairs used to be the bread-and-butter of a Computer Support department. Upgrades beyond those involving RAM memory are less popular as well, as apparent by the increasingly modularized and integrated designs of computers. Options that were traditionally add-on components, such as video and sound, and network connectivity, and modems are now commonplace, integrated into the majority of computer systems, precluding the easy upgrade or replacement of these components.

I don't think that this trend is bad, although many computer owners' philosophies may differ from mine. My opinion is that a computer was designed and engineered to work well given a certain range of variables, such as the following:

Application *Brand*

Application *Generation*

Operating System *Brand*

Operating System *Generation*

Hard Disk Size and *Generation*

Peripheral Selection and *Generation*

Leaving the range of these variables can, in my opinion, compromise the overall operability of your systems. It is in this thinking that I wrote the section about client PCs. Likewise, I think that shoveling equipment onto your computer network's backbone is a bad plan, and will have an adverse effect on *everyone* who uses the network.

So, what does Obsolescence Management mean? I think that it means not to put brand new equipment into three-or five-year-old computers. I try to stockpile enough parts of a correct generation of equipment so that there are no surprises when one opens the case. Finding some kind of brand new device in an old PC is usually disheartening for any computer repair tech. In addition, keeping the drivers and books for equipment until it is retired is a good idea. Companies are sold, manufacturers change product support options. Having the original disks and manuals never hurt anyone.

Support has improved dramatically in some ways due to the Internet. Provided that you have an Internet connection, software and documentation for your equipment is many times only a few clicks away.

Bear in mind, though, that part of obsolescence management is knowing when to ditch the old equipment. Computer equipment that cannot be readily or cheaply repaired, due to age, should be gracefully retired and replaced.

Looking back to the budgeting section, you will see the following section:

V. PROACTIVE MAINTENANCE ITEMS

=-=-=-=-=-=-=-=-=-=-=

A. 10 PERSONAL COMPUTERS   11.1%
To replace old equipment in offices which is still in use

Good proactive maintenance is to perpetuate the revolving door of technology. If you upgrade a certain percentage, perhaps 10%, of your installation you can avoid having to upgrade a significant percentage of equipment later. This is easier on your budget, and easier on you as the person doing the upgrades.

Never allow computer equipment or software to age excessively in your organization. Key systems which are no longer supported by their manufacturers, or are more than two generations old are ticking time bombs.

Once upon a time, I worked as a contractor for a large, established bank in Baltimore. One of the things we supported was a large volume of IBM OS/2-based Equipment. OS/2 is a commercial grade operating system, sim-

ilar to Windows NT. Much of the computer world has moved beyond OS/2, and support is no longer as comprehensive as it once was in the past.

One of the OS/2 computers ran a particular program that got daily mutual fund pricing information. It then propagated this information to other custom programs for further analysis. One day it stopped working, and we had to reboot this computer.

None of us had rebooted this computer as long as I had worked there. It was on the battery backup system, and sat in one of the system rooms. Nobody ever touched it, and it didn't seem sexy or important. It just worked, and then one day it didn't.

We rebooted it, thinking that it would be a simple repair. The video card, however, had different ideas. It wasn't just any video card, it was a proprietary IBM video card which required something called a Reference Disk to work properly.

Now, technically speaking, OS/2 had done nothing wrong. The problem was that as the battery had died, we could not get the computer to work with the video card, so we removed the card. The computer would not boot-up at all, asking now for the computer's setup disk.

Nobody working there had ever seen the documents for either the computer (which itself was from the very early 1990s) or the video card. No documentation was to be found, and thus no daily fund pricing. It turns out that this was a lot more important to a great many more people than we thought. Because the computer was so established, shall we say, many groups of people had latched onto it when writing custom programs, none of which were working.

There is a moral to this story, and it isn't necessarily about keeping the diskettes. This computer should have been upgraded and replaced like most everything else, instead of deported to the back of a system room with a bunch of other obsolete, marginally working junk which nobody had the, uh, gumption to replace.

LEGACY SYSTEMS AND DATA CONVERSION

Possibly the most aggravating aspect of management is dealing with the legacy of something old. The culture surrounding legacy systems always seems to be the same. There are two types of killers when it comes to upgrading a system:

The first problem is Resistance to Change with regard to the old system and practices. Lack of buy-in issues may have to do with feelings of loss of control or power, or a resistance to spending money to upgrade a system.

The second problem is much more sinister, because it seems innocuous. This problem is Looking at the new system as a panacea, a universal elixir to any and all problems. "This email system will fix all of your ills". This happens when you start believing the sales pitch, and become the salesman in your organization. Problems in this area are much worse for you, because they work against the natural fact that every system has limitations. This may not always be the case, but for the foreseeable future, there will be limitations in any kind of data oriented system. We are at that evolutionary stage.

There are certain givens of legacy systems in small organizations. If it seems that these are somewhat cynical, it may be. I have seen many small systems, and these four rules are certainly proven, empirically.

Rule Number 1: Nobody documented the old system, which was installed by the (A) Old system manager, or (B) an old, out-of-business contractor.

Rule Number 2: There is something really quirky or irritating about the old system, which nobody can explain or fix, and everyone else has gotten used to. "Oh, that, well it crashes every day, you just have to remember to restart it in the morning, it's no big deal. See, we even have here a rotisserie list of whose job it is to do".

Rule Number 3: Something about the old system is either completely proprietary or completely outdated. Data cannot be readily retrieved or converted to a new system.

Rule Number 4: There exists some amount of uncertainty about switching away from something old to something new. The old "Devil you know is better than the devil you don't" adage. This is possibly the biggest stumbling block, because it affects resistance and buy-In to the idea of a new system.

Communication and training can counteract these problems, and can work to build a culture of willingness to try new ideas.

Data Conversion

Possibly, data conversion from a legacy system is the most important stage when planning a revitalization effort. Unlike a new install, there is a great deal of baggage hanging around from the old system. This baggage is in the form of practices and data.

Practices can be overhauled. Most reasonable people will balk about the move to something new. Up-front and honest communication, and training early and often, especially about the changes, will help this effort.

Data transfer is more difficult. Although data is stored in standard forms within a data system, The structure of data is usually stored in a proprietary format within all but the simplest programs. Databases, especially, are rarely designed around any standards. This makes software a "one-way door", which is what the manufacturers of software are trying to do, indirectly. Advanced features which separate software, will often preclude certain features of data to be converted. It is important to be aware of any shortcomings up front, so that they can be dealt with prior to the conversion.

In most cases, try to arrange an agreement with the vendor of the new system to do the data conversion for you. System integrators are more experienced than you are in data conversion. Making the integrator responsible for data conversion can save you headaches during the already busy process of an upgrade. Sometimes, data conversion can be tricky, especially if there is no defined data upgrade path. Certain software has a path to it from other programs. Depending on your software, there may or may not be a path.

This is a key reason why you should not let systems age too much. Usually, software can convert from recent versions of other programs. The key part of the last sentence is *recent versions*. A program written in the early 90's is not recent. Although conversion may be possible, there may be hang-ups or compromises that must be made. To help with this, sometimes data can be converted to an intermediary program in order to be converted to the new program. This too can be tricky or expensive, as it requires an additional conversion to a piece of software that the system integrator may not be as familiar with.

Practices also become involved here. Data is sometimes not so straightforward. In the case of certain programs, the data or information that is generated is the advantage of a program. Systems are usually designed initially with goals in mind, or features that work for a particular organization. Reports may be generated that cannot be replicated exactly in the new software.

Look at the Original Goals

Take a look at the original goals, if they were written down. If not, try to find the people involved with the original system. It is very helpful to see why the decisions were made to go with a system in the first place. Data analysis programs, especially, are chosen because of their flexibility or their ability to generate certain types of reports or their performance of specific functionality.

CROSS-PLATFORM ISSUES

One major stumbling block when dealing with legacy equipment is making the move from one standard to a newer or different standard.

Sometimes, this does not matter, as with moving from one brand of Windows computer to another brand of Windows computer. It is likely that the old software and information can be transferred to the new machine without a great deal of difficulty.

Cross-platform issues usually manifest in one of two ways. Either a server or program will no longer operate on a new version of the client's operating system, or a move has been made to a completely different operating environment, and no programs can be readily transferred.

For instance, consider the first type of situation. Your organization used an old version of DOS, and was historically entirely a DOS-based operation, with DOS-based programs. The organization decided to move to new machines, and to Windows, as many organizations do. There is a clear upgrade path between DOS and Windows, as the platforms are similar, and your older programs will likely continue to operate in the new environment.

In this case, it may be necessary only to move the key programs and data from the old computer to the new, which may be as simple as saving the information to the network, and moving it back to the new computer after it is installed.

The only stumbling point here is if the older program will not work in the new environment, or will not work in a satisfactory way. Many times this does occur. In this situation, systems may need to be modified to handle the older software, or the data may need to be converted to run in some new alternative to the older program.

A more complex situation may occur if, as in the latter situation, you move to a different platform entirely. This occurs in organizations where the switch has been made between Windows/DOS and the Macintosh system. There are software packages on both sides of the street that allow DOS files and disks to be accessed on the Mac, and vice-versa.

Conversion is not easy, as is access to the server standards. It is difficult, although not impossible, to get the Mac to recognize the PC servers, and the PCs to recognize the Mac servers. There are many independent software companies producing packages which can be used to aid in this process.

In both of these situations, tantamount to the actual conversion of the data is the training issues and expectations surrounding the new software. Data can usually be converted, or accessed in one way or another on the new system. Managing the expectations and proper use of the new software, especially in the light of prior experience with the old software, is more of an issue. Most software packages vary from older versions, or with regard to competing products.

Training And Conversion

Training is the key to managing expectations. On one hand, there is a great advantage, as the users of the system are, by and large, comfortable with the process as an electronic process. On the other hand, they may be used to the medium and manners of the old system. Some of your users may complain about the switch, which usually manifests as "I liked the way the old program did this better!" Some users may resist the change. Address these concerns with knowledge and sensitivity. Look at the needs documents, and make sure that the program does the same kinds of things as the old software, because key features that turn up missing can be a real headache.

The key thing about conversion to new systems is that it is inevitable. New technologies come, old ones go. Renewal is part of growth. If renewal wasn't important, there would be no need for technology, for this book, or for any of this.

BUSINESS CONTINUATION

If you take away one idea from this book, it should be this: Maintaining a data system in an organization is a lot like any other business continuation service, like facilities management, contract management, accounting, or filing. These services all have things in common. These services are necessary to operating the business side of your non-profit organization. They, by nature, supercede the people doing the functions. Documentation and functional backup are ways to ensure that the processes and infrastructure continue to operate, even in the absence of those who do the jobs. If duplicity is the goal, the organization should avoid situations where they are reliant on one person to perform a function. For instance, the operation of the accounting or database systems should be known by more than one person.

This is a fine line in most small businesses, because people, by nature, will become the resources for these applications. You may have a secretary who "is the guru" of the database program, or the phone system. Because of the level of interaction, your primary accounting person will have more knowledge of the accounting system than the people who maintain the computer systems at large.

People, especially volunteers, move on. Life takes us all in different directions. Some are predictable and can be planned, others are just as unpredictable. Maintaining documentation, such as maps of your facilities, lists of numbers for contact persons, and the like can make a transition much easier. In addition, solid membership in your technology committee can make all of the difference, and can help to bridge gaps in business services.

CONCLUSIONS

There really isn't a whole lot more to say. This is a good thing, because this book is quickly coming to a close.

Remember that this is a guide, and very much a first step in a long process for all involved. Before you make any decisions, You need to find books which actually discuss hard concepts, like network design and hardware. You know what I mean, those books which list pages programming examples, and are a thousand pages long. Many of the topics have been discussed, whether they be technical concepts, or management techniques are pretty standard. There are lots of good books out there that discuss these concepts in detail.

Before you retire this book to the shelf, and maybe the trashcan, take a look at the Appendices, and especially the "Resources" pages. Obviously, I am not specifically endorsing everything everyone says in these books, but they are generally sound places to start, and they are written well. To be a writer, one must love to read. I love to read, and these books, and all hold a special place in my heart.

So, having no more to say, I leave you on your journey. Good luck and God Bless.

A. RESOURCES

There are many good books, as well as other resources, which describe topics dealt with in this book. These are books that I have read, and can personally recommend, not some generic titles that show up in some search engine. Specific network technologies change rapidly. In an effort to maintain the significance of this book, I have tried to paint broad strokes, rather than to focus on specific technologies or brand names.

I do feel, however, that there is a place for more technical volumes in anyone's library. This book is a stepping stone, hopefully one of your first in this process. Its purpose is to lead you in a direction. The books listed below may help you in the next step, which is to begin fleshing in your plan, and to determine what technologies are appropriate for your installation.

I hope that you find them as useful as I have.

(1) Fire in the Valley
This is an excellent book about the first ten years or so of personal computing. This book is a real "page turner", a rarity amongst computer books.

Paul Freiberger and Michael Swaine: Fire in the Valley: The Making of the Personal Computer. McGraw-Hill Company, USA 2000
ISBN: 0-07-135892-7

(2) System Analysis and Design
A textbook, this is a good starting point for both the SDLC and topics such as data-flow diagramming.

Shin Yen Wu a Margaret S. Wu: System Analysis and Design. WEST PUBLISHING COMPANY, USA 1994.

(3) Managing a Windows NT Network: Notes from the Field
A good book, if you can find it. This was a volume published by Microsoft regarding people in the field using Windows NT. Has a lot of scripting and network setup examples. It does, however, focus on larger systems and Wide-Area (multiple location) networks.

Microsoft Corporation: Managing a Windows NT Network: Notes from the Field. Microsoft Press, Redmond, WA USA 1999
ISBN: 0-7356-0647-1

Practical Network Cabling
A great handbook about network wiring

Frank Derfler and Les Freed: Practical Network Cabling. Que, a division of MacMillan Press USA, USA 2000
ISBN: 0-7897-2247-x

Data and Telecommunications: Third Edition
This book discusses everything and anything about network cabling and telecommunications (the phones)

Regis J. Bates and Donald W. Gregory: Voice & Data Communications Handbook.McGraw-Hill, USA 2000
ISBN: 0-07-212276-5

How to Manage the I. T. Helpdesk
A good book dealing with issues having to do with managing a helpdesk.. This book discusses lots of 'soft' issues like support and training.

Noel Bruton: How to Manage the I. T. Helpdesk. Butterworth-Heinemann
May 1997
ISBN: 0750638117

B. LISTING OF PICTURES, DIAGRAMS AND TABLES

C. CASES

The following are cases that may help to illustrate a few concrete examples of common types of networks.

CASE 1: SHARED DIAL-UP INTERNET ACCESS MODEL

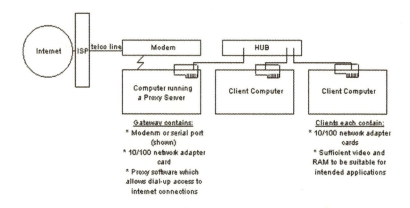

FIGURE 5.1: SHARED DIAL-UP INTERNET ACCESS MODEL

The shared dial-up Internet access model is usually the first iteration of an organization's Internet connectivity. It usually involves a simple network, in this case of the star configuration. This network may include two to ten PCs, as this is the outset of the reasonable number of computers which can function reliably and adequetly under this arrangement.

Each client PC, in addition to its regular compliment of hardware, includes a network adapter card which supports TCP/IP. TCP/IP is the only workable network protocol in this situation. On one PC dedicated to this purpose, a piece of software running something called a *Proxy Server* is installed on the network. This computer may have either an internal (card-type) modem, or an external serial modem. The modem is connected to a telephone line, and the proxy server software is configured to dial a predetermined telephone number to the Internet Service Provider.

Despite what some ISPs may lead you to believe, this is a legitimate use of one dial-up account. Your total use, whether you have 1 PC or

10 PCs connected to this network, is a hypothetical 56Kbytes, the maximum speed available to this type of arrangement. This is the same amount of hypothetical speed available to one PC connected to the Internet via a dial-up modem.

PROS

- Easy to set up

- Inexpensive

- Proxy server can be reusable if a different type of Internet connection is used

- A very safe connection which can negate the need for a firewall

CONS

- Slow

- Proxy servers are notoriously incompatible with many types of Internet content, (eg. Realplayer, telnet)

CASE 2: SHARED DIAL-UP INTERNET ACCESS MODEL WITH MODEM POOLING

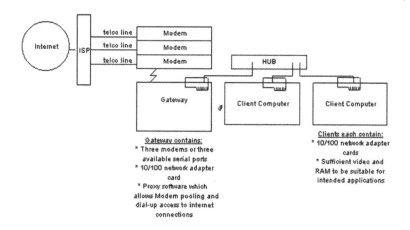

FIGURE 5.2: POOLED DIAL-UP MODEM MODEL

This arrangement is an advancement of the first type of dial-up connection network. In this situation, the proxy server is configured with either

- Multiple internal Card-type modems, each configured with separate dial-up account

- Multiple serial cards, attached to external serial modems, each configured with a separate dial-up account

How does it work?

On the computer running the Proxy Server software component (Called the "Gateway" computer), the proxy server is supplemented by another piece of software providing a service called "modem pooling". Modem pooling is available by inclusive service programs in certain recent versions of Microsoft® Windows™. Dial-up access to the

Internet is established by simultaneously using two to ten modems, each having its own separate telephone line. The hypothetical speed of the Internet connection can be calculated by the following:

NUMBER OF MODEMS *X* **56** = total speed in Kilobytes per second

Prudence dictates that you subtract 25% of this number to arrive at your actual speed, due to factors outside of our control. Modem pooling may have a slight adverse speed affect. In addition, due to line noise and distances, modems do not achieve their maximum speeds.

CASE 3: A ROUTED DSL (OR CABLE) INTERNET CONNECTION USING A FIREWALL

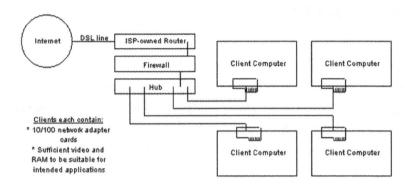

FIGURE 5.3: A ROUTED DSL/CABLE-MODEM MODEL

This is another popular network configuration. This network plan could be used if you want to go with a higher-speed network, an "always connected" configuration. With today's high-speed Internet access, it is important to have either a software or hardware based firewall product in-line between your ISP's communications equipment and the local hub.

The network clients in this example contain network cards and are configured with TCP/IP which connect them to the hub and allow them to communicate with the Internet. The hub, in turn, is connected to the firewall. The firewall in this example would have a TCP/IP DHCP (address-control) server installed, which would control network addressing and provide routing functions to the Internet.

This network configuration is very scalable, allowing access of two to hundreds of network clients.

CASE 4: A SINGLE-SERVER NETWORK

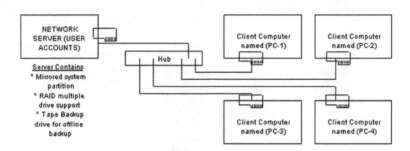

FIGURE 5.4: A SINGLE-SERVER NETWORK

The single-server network is basically a group of PCs which have network adapter cards installed. These computers are in turn, connected to a hub, and to a single, or multiple servers which are all located within close proximity to the client computers.

This type of network is a good choice for sharing of information, or centralized data backup. This model can is used in installations with 2 computers, but can readily be scaled up to networks with dozens of client computers by the addition of additional hubs.

CASE 5: A SWITCHED NETWORK WITH FIBER OPTIC BACKBONE

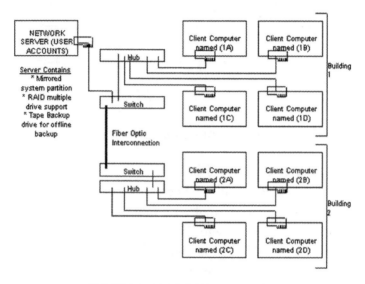

FIGURE 5.5: A MULTI-ZONE NETWORK

A natural progression of any network is to a multi-zone network with a backbone. These types of networks are used in situations where multiple buildings are connected to the same network, or where a building has multiple floors, in which case the backbone may also be referred to as a riser cable. Depending on the age of the system, a network's backbone may be a fiber optic cable, copper cable, or coaxial cable.

Fiber-optic based backbones also require a network switch equipped with a fiber transceiver. For each interconnection, Fiber optic requires a receiving and sending line, that is 1 pair of fiber lines, and a transceiver on each end of the connection.

When multiple wiring closets are employed, a two-level star topology is employed. There is a meta star, whereby a main closet exists (usually in the primary system room) is called the "MDF", the Main

Data Frame. The branches of the star are subsidiary wire closets on each floor or in the other buildings are called "Id's", Intermediary Data Frame(s). The MDF connects to all of the IDFs via the backbone connections, then each IDF connects in a star-topology.

The below figure on the left shows the logical design of a star network, the figure on the right shows the physical design of a star network.

D. A BASIC NETWORK ADMINISTRATOR'S TOOL KIT

Because most network people are basically hardware guys, and I am no exception, I could not resist putting something in about a network tool box. Every network Admin should have a tool kit, if for nothing else than to honor the essence of what an Administrator's job really is, the "Network Repairman". Unlike the affable and usually idle "Maytag Repairman", the stereotypical Network Repairman may smoke *a lot*, maintains an irritated scowl most of the time, and may be prone to throwing books or other convenient objects at others who ask stupid questions about the system. (Just Kidding)

Network administrators are not expected to be installers, and need not maintain a complete set of installers' tools. However, with knowledge and experience comes power. Power in this case may be the ability to diagnose or fix something yourself in lieu of having to call a contractor.

SECTION 1: TOOLS

A GOOD, STURDY, TOOL BAG: Boxes are hard to carry, notebook computer bags are not sturdy. Get a good electricians bag, one with a shoulder strap. Many are made to fit in with a professional look, as well as a more traditional electricians look. In lieu of this, a sturdy canvas or nylon bookbag may be used.

STANDARDS HANDBOOK: A good EIA/TIA reference standards book will aid you in the proper installation and testing of your hardware.

FISH TAPE: Is a large wound spool of somewhat rigid metal wire. Sometimes, it is encased in a plastic housing, which makes it easier to store and manage. Fish Tape is used to aid in pulling, or sometimes pushing, wire around in ceilings and through walls. It comes in spools of anywhere from 10ft to 30ft. You need a good fish tape if you are going to pull wire. In effect, one is essential if you are going to be able to add data jacks in a building. The good 20ft spool in a plastic dispenser should suffice for most jobs.

PUNCH DOWN TOOL: This tool is used to attach the wires on the server-room side of the network. It literally "punches down" each wire to the terminal on the patch panel in the wire closet and on the other end, to attach the wire to the data jack in the wall. Punch down tools are useful for repairing data jacks, as well as installing new jacks.

CRIMP TOOL: The crimp tool is used to make patch cables. Patch cables look like the cable between your telephone and the jack in the wall. The crimp tool compresses the plastic fitting around the wire, so that the plastic jack will be functional. Telephone jacks use a jack called the RJ-11 standard. Networks use a jack called the RJ-45. The RJ-45 has 8 wires, and is larger in size than the telephone jack. Usually, you would want to get a crimp tool which can crimp either size. These tools have two sizes of openings, one for the smaller RJ-11 and one for the larger RJ-45 types of jacks.

WIRE CUTTERS: A good set of wire cutters and strippers is handy in many situations.

WIRE PREPARATION TOOL FOR STP: Shielded Twisted Pair wiring (used in network applications) needs to be cut and stripped prior to use. Special wire preparation tools with blades fixed at certain distances from the wire make preparation very easy.

WIRE END-TO-END TESTER: An end-to-end tester is a basic electronic tool which tests a length of wire. This length can be a short interconnection cable, or can be a long cable running through a building. There are two boxes, one attaches to each side of the wire. Using an array of lights, these tests show the suitability of the wire and the quality of the installation job. There are an array of other testers for network wiring, but the end-to-end testers are the least expensive, and most versitile for occasional use.

AWL AND DRYWALL SAW: You know what an awl is. A drywall saw is a small hand saw used to make small cuts in drywall. Check out a home repair book for instructions on how to make holes and cuts in drywall.

AN ELECTRICIANS' LEVEL: This is a small, lightweight level which is convenient when setting up a new data outlet.

SCREWDRIVERS: Phillips-head, slotted, torx, the more sizes and types the merrier.

A TELEPHONE TEST SET, AKA A "BUTT SET": The Butt Set is is historically the first piece of equipment that a telephone service person or installer would have on their person. It looks like a heavy-duty telephone handset, complete with a dial-pad, and a range of other features are available on these units. Instead of having a modular plug on the end, however, the wire coming out of the telephone attaches to two modified "alligator clips", which allow you to connect to the premise wires for the telephone system. In this day and age of a line in the sand between your premise equipment, and the wire that the phone company is responsible for, this piece of equipment is invaluable to helping the Telephone company determine where the problem lies.

PART 2: PARTS and MISCELLANEOUS

RJ-45 ENDS: These are the clear plastic ends which are used with the crimp tool. The RJ-45 standard type of end is used most often with modern networks.

BOX OF CAT5 WIRE: A box of CAT5 or CAT5e wire, depending on what is installed in your building. This is necessary for any repair or addition project. Boxes usually come with 200 or 400 ft of wire, spooled in the box. This makes installation, and pulling of wire, easy.

KEYSTONE DATA JACKS: These are the actual jacks to which you attach the wire at the user's end. These jacks are rated to the specifications that your network conforms to, and are designed to fit into the plastic faceplace which is attached to the wall. Many manufacturers make data jacks. Usually, you want to keep the data jack and the faceplate the same brand.

FACE PLATES AND INSTALLATION HARDWARE: The faceplate is the visible plastic part which covers the hole in the wall. While all wall applications call for the same standard faceplate, There are different applications of installation hardware (the part that holds the plate to the wall) for different types of walls. A block wall is different from a drywall, or a wooden wallboard. Check with your local electrical or data equipment supplier to determine what is appropriate for your type of installation.

WIRE MANAGEMENT TIES: Wire management ties are the duct tape of the network. You can use these plastic ties to hold everything from wire, to power strips, to holding the racks together. Their varied uses could fill a book on their own. You should have many sizes and lengths of these ties.

ELECTRICAL TAPE: Great for temporarily holding together of wires, colors can be used to label things. Get as many colors as you can, and always keep some on hand.

RACK MOUNT AND PC SCREWS: Depending on how much work you do, you should always have some extra PC screws. There are only a few types of screws, for the case, fan, and internal components. Little bags of screws are available at most computer and electronics stores.

PART 3: SOFTWARE and COMPUTER HARDWARE

OPERATING SYSTEM DISKS/CDs: I always make a note to carry around a copy of the operating system disks for each OS that is in use. This makes additions of peripheral equipment or repair to the operating system easier.

STARTUP DISK WITH CD-ROM DRIVERS: As with the Operating System CD, the startup disk is essential, especially when you get to that machine that just will not boot up.

DISK UTILITIES SOFTWARE: There are a myriad of disk utilities programs available on the market. Carrying a disk with one of these programs on it is not a bad idea.

ANTIVIRUS PROGRAM DISK: As with the disk utilities software, antivirus disks should be carried. This may aid in troubleshooting potential virus problems.

A CROSSOVER CABLE: If you deal in an Ethernet network, as most everyone does, a crossover cable is a handy thing to have. These cables are readily available through computer stores and catalogues. Their purpose is to connect two computers, or ethernet devices together, without any additional equipment, such as a hub. If you are trying to move information off of a computer, or are trying to test a network card, it is an invaluable tool.

A FEW BLANK FLOPPY DISKS: Just in Case.

APPENDIX A

OUTLINE FOR A NETWORK EMERGENCY PLAN

In 2001, we decided that there was a necessity for a documentation project which would outline the performance of key functions, and illustrate a layout for the network. It was called, playfully, the "MACK Truck" Project, under the pretense that if the system manager was to be hit by one of the many "MACK trucks" which cross the public street which divides our campus, that someone else would need key information regarding the structure of our system.

SECTION 1: NETWORK GENERICS

CHAPTER 1: Our Network

> Overview
>
> Physical Attributes
>
> Logical Attributes
>
> Hardware (Server)
>
> Software (Server)
>
> Hardware (Client)
>
> Software (Client)
>
> Network Protocols

Daily Check of the Mail Server logs

Daily Check of the Tape Backup Server logs

Weekly cleaning of the "administrator" email account

Weekly preventative maintenance of the Tape Backup Drives

Weekly preventative maintenance of the Voice Mail Computer

Weekly preventative maintenance of the Computer Lab Equipment

Irregular Maintenance OVERVIEW

Maintenance of Printers

Changing of toner cartridges

If a printer breaks

If a desktop computer breaks

Reinstalling Office or WordPerfect should the need arise

Reinstalling Email on a computer which has forgotten it's password or username

What if someone wants to check their email from home?

Problems with Powerschool

Problems with Great Plains Accounting Software

CHAPTER 5: MORE SERIOUS ISSUES

What if a serious problem occurs

A Framework for troubleshooting problems with the computer system

Being calm in the face of a problem

A Framework for troubleshooting problems with the servers

Printing

SECTION 2: NETWORK SPECIFICS

CHAPTER 7: THE MAIL SERVER

Specifics

When is it down

What are ramifications of it's being down?

Troubleshooting

Restarting the server

A note on email services

CHAPTER 8: THE PRINTER AND INTRANET SERVER

Specifics

When is it down

What are ramifications of it's being down?

PRINTER RELATED ISSUES

PRINTER DIALOGUE BOXES

PAPER

STOPPING AND RESTARTING THE QUEUES

INTRANET RELATED ISSUES

MONITORING IIS

STOPPING AND RESTARTING IIS

Restarting the server

CHAPTER 9: PDC/BDC SERVERS

Specifics

When are they down

What are ramifications of their being down?

Troubleshooting

Restarting the server

CHAPTER 10: BACKUP SERVERS

Specifics

When is it down

What are ramifications of it's being down?

Troubleshooting

Restarting the server

CHAPTER 11: CD-ROM SERVERS

Specifics

When is it down

What are ramifications of it's being down?

Restarting the server

CHAPTER 12: INTERNET ACCESS

When is it down

What are ramifications of it's being down?

Restarting the covad router

Restarting our router

CHAPTER 13: SWITCHES AND HUBS

When are they down

Restarting this equipment

CHAPTER 14: THE COMPUTER LAB

Electrical Service to the Lab

The Hubs

The Fan in the ceiling

CHAPTER 15: CONTACTING OUR VENDORS

PC SUPPLIES/NETWORK SUPPLIES/PRINTER SUP-PLIES

PC PARTS/REPAIR OF PRINTERS/REPAIR OF DESK-TOP PCs

HP PRINTER SUPPORT

NETWORK DEVICE SUPPORT

SECTION 3: NON-NETWORK ITEMS

CHAPTER 16: THE TELEPHONE SWITCHING SYSTEM

Specifics

When is it down

Ramifications of it being down

BEFORE YOU RESET THIS DEVICE

Rebooting the phone switch

Monitoring the Phone Switch

CHAPTER 17: THE VOICE MAIL SYSTEM

Specifics

When is it down

Ramifications of it being down

System Maintenance

System Reboots

CHAPTER 18: THE SOUND AND VIDEO SYSTEMS

The sound system keys

The microphones and AV equipment

Calling The Sound Contractor to resolve problems

Calling to resolve Cable TV or video system problems

CHAPTER 19: THE ALARM SYSTEMS

The Fire Alarm

The Burglar Alarm

Calling The Alarm Monitoring Company about Alarm Problems

Calling Alarm Monitoring if an alarm has triggered

APPENDICES

1. **Plant Physical Layout showing IDF and wire closets**

2. **Server Room Overview**

3. **Network Layout**

4. **Tape Backup Procedures**

5. **Cleaning the Mail Server Account**

6. **Lab PM documents**

7. **Contact List**

8. **Quick List of Phone numbers and Parish Telephone Numbers**

APPENDIX B

SERVER OUTAGE SECTION OF A NETWORK EMERGENCY PLAN

I decided to include a chapter out of a recent internal document that I created for our organization. Some form of emergency manual is necessary because it lets the remaining staff know how to deal with issues that may arise, and some suggestions on who to call to resolve them.

Some of the issues at the beginning are normal maintenance items, others are abnormal, and honestly, most non-technical people are not going to be able to fix these problems. Basically, the documentation gives information, for these instances, on how to get in touch with the appropriate contractors or vendors.

I decided to break this into two sections. The first is chapter 5 from the network emergency plan. This is a basic framework for dealing with some basic problems.

The second section is a sheet that I made for each server. This is a sheet which shows what this server does, how you know it's down, and what to do about it.

REGULAR MAINTENANCE

The network requires a certain amount of baby-sitting items. Among them are:

- Daily Backup Tape rotation
- Daily Checking the event log on the Print Server
- Daily Checking the event log on the Mail Server
- Daily Checking the event log on the Tape Backup Servers
- Weekly Cleaning of the administrator email account
- Weekly Preventative Maintenance of the Tape Backup Drives
- Weekly Preventative Maintenance of the Voice Mail Computer
- Weekly Preventative Maintenance of the Computer Lab

There are other minor items which may need to be addressed, irregularly:

- Maintenance of the Printers
- Changing of toner cartridges
- If a printer breaks
- If a desktop computer breaks
- Reinstating Email on a computer which has forgotten its password
- Irregular problems with Powerschool and/or powergrade which do not have to do with the Internet connection

- What if the parish database system fails

- what if Great Plains fails

Tape Backup Rotation

There are two sets of backup tapes. The tapes must be changed daily. See the appendix for specific instructions

Daily Checking the event log on the Print Server

The print server monitor is on the right hand side, under the computer marked "network".

1. Turn on the display if it is not turned on already

2. Log in if necessary

3. Under the <u>Start</u> menu at the bottom of the screen, choose <u>Programs</u>, then <u>Administrative Tools</u>, then "EVENT VIEWER".

The event viewer window should appear on the screen.

The Event Viewer will show all of the things which have been printed by people using the network. You are not looking for anything in particular. Problems which manifest themselves, however, usually start by showing error messages in the event viewer. A lot of messages to the event viewer are normal, and even many of the "medium level" warnings are normal, but you do not want to see very many "serious" (red) messages from the print server.

Daily Checking the event log on the Mail Server

The mail server is on the right hand side, if you are facing the equipment rack.

The event viewer for the mail program is on the screen of the mail server. The monitor is connected to several other servers as well as the mail server. If the monitor is not showing the Mail Server display, See the instructions in Chapter 5 about operation of the KDM switch.

The event viewer for the mail server is on the screen of the mail server. Look for connection error messages. This will indicate that some type of error is occurring with the transfer of mail. Look at the top of the event viewer window. There should be "0" outgoing messages waiting to send. BUT, there should be lots*hundreds*of incoming messages waiting.

Checking this server is especially important, because due to the nature of the mail server operation, problems can occur which do not get reported for a day or so.

Daily Checking the event log on the Tape Backup Server

Checking the event log on the Tape Backup server is easy. Simply turn on the display next to the 8track computer on the right hand side. The tape backup monitor is already loaded.

Look for "FAILURE" status readings. This indicates that the previous day's backup did not run successfully. This, in all likelihood, means a problem with that day's backup tape.

Weekly Cleaning of the administrator email account

The administrator account on the email server is used for only one purpose. Log files and "lost" email files show up here. It fills up rapidly, and needs to be emptied on a regular basis.

See appendix 5 for instructions about how to do this.

Weekly Preventative Maintenance of the Tape Backup Drives

On Mondays of every week, the tape backup drive needs to be cleaned with the "CLEANING" tape. Between removal of the previous Friday's tape, and the insertion of the next Monday tape, the Cleaning tape must be:

Marked twice—once for the first drive
              —once for the second drive

and then inserted into the first drive

and then inserted into the second drive

Then, returned to the tape storage box.

Finally, you must mark an "OK" on the "tape PM" calendar on the wall outside the system room.

Weekly Preventative Maintenance of the Voice Mail Computer

The voice mail computer needs to be restarted weekly. I do this after 5 PM on Tuesdays, because it causes a momentary lapse in voice mail access for outgoing and incoming calls.

Periodically, MS-DOS defragmentation must also be run on this computer.

Weekly Preventative Maintenance of the Computer Lab

The computer lab is to be visited weekly on Mondays to determine that all computers and printers are functional

Functionality is defined as:

> Able to print
>
> Sound is OK
>
> Network connectivity is working
>
> Desktop arrangement is as documented

IRREGULAR MAINTENANCE

Maintenance of Printers

Periodically, printers jam. The manuals and documentation for our printers are available on the Manufacturers website
If the problem calls for more serious attention, our newer printers have on-site service contracts through March, 2002. These service contracts cover repairs to these printers.
The documentation is in the "service contract" file in the network administrator's office

Changing of toner cartridges

Toner cartridges are available in the Network Administrator's office. If more toner needs to be ordered, it can be ordered through the mail order vendor

[MAIL ORDER VENDOR INFORMATION]

If a printer breaks
If a printer breaks, see documentation in the "network administrator's" office under the "Service Contract" file about getting the printer fixed.

We contract also with a gentleman who can come on site to do printer repair on HP printers.

[PRINTER REPAIR CONTRACTOR'S NAME AND NUMBER]

If a desktop computer breaks

If a desktop computer fails, take it to either:

(LOCAL PC VENDOR WHOM WE DEAL WITH]

or

[ALTERNATE LOCAL VENDOR]

We have accounts for parts and systems at both locations.

[ACCOUNT INFORMATION]

Reinstating Email on a computer which has forgotten its password
Periodically, the outlook express email client will "forget" the email
password.

The email password for everyone here is the same as that on their username.

What if the Parish Database Application fails?

If the database application fails, we have a service contract with the
manufacturer. The office has the telephone number in the list of con-
tacts.

What if there is a problem with the accounting software.

If the accounting software fails, you may have to call the great plains
support company, called the consulting group out of Baltimore.
[CONTRACTOR NUMBER AND INFO]

More Serious Issues

What if there is a more serious problem?

A more serious problem, around here, is a problem which is affecting
more than one user, because of a system outage. Unfortunately, many
users think that there is a system outage, when there really isn't.

A framework for troubleshooting Problems with the computer system

A person may report a problem, i.e.:

"The email server is down, I cannot get my email", or "I cannot print"

You will want to use this methodology:

1. Ask the user to restart his or her computer, login, and go back into the program. Even if they "have done this already", make sure that they do it again. Restarting corrects a surprising variety of problems in Windows.

2. **Check the service.**
    Does the email program seem to be working correctly? Is the program crashing or giving an error code or message when started?

3. If the program is working, or seems to be working, Can the user send email?, Receive email?
    send a message to yourself, then send a message to that user. Many times, the email problem has nothing to do with our network, but is on the "other end".

    Is the person sending mail to a valid email address?

4. Is there an error message when trying to send/receive mail?

5. Is there an error message being generated by the operating system

6. Is the mail server working?

    Are you able to get email/send email?

    Are others complaining about this issue?

7. If you have determined a server problem, check the status window for that server—in this case, MAIL.
    If our Internet connection has failed, this will let you know that there is an outage. If there is a server problem like an error message on the screen of the server, try restarting the server.

Being Calm in the face of a problem

Unfortunately, you may encounter a problem with one or more of the servers during your tenure as Network Trustee. While the servers are reliable, they will fail occasionally.

The key to managing server problems is simple: Be Calm. Even if others are agitated or displeased by an outage, you need to keep calm during a problem. Remember: There really are NO computer emergencies. The mail will eventually go through, the phone system will eventually be restored.

Communication
The key to keeping order and calm in the face of a problem, I have found, is communicating frequently with key people here about the current status.

If possible, I have in the past sent email to the secretaries, or if not possible, called on the phone. Many of the staff will call [The Technology Department Secretary]. Faculty will call [The school secretary]. It is important to keep communications open with the secretaries, especially with regard to telephone related problems, as they are the first line of defense on this front.

If possible, in the case of a telephone outage, make an "all call" through the system, to allow others to know that an outage is occurring.

A generic framework for fixing Server Problems

You may discover that the problem lies in the server. There will be, in subsequent chapters, more in-depth information about each server, but here is a basic framework for the servers here.

1. Check the server using the KDM switch

The KDM Switch

The KDM (keyboard display monitor) switch is a box which allows up to four computers to be controlled by a single mouse/keyboard and monitor.

Pressing the "channel" button on the KDM switches between these computers. The only one logged in is Mail, which is logged in under the ADMINISTRATOR account.

MAIL has a green background, and a status window on the screen.

2. Look at the status window

Check the Event Viewer, under Windows NT. On the mail server, look at the status window for the Mail Program.

Look for obvious problems, like "Windows Protection Error" messages, or the "Blue Screen". Look for crashed servers. Sometimes this will happen.
If you find that the server has crashed, Restart it.

What if the server does not appear to have crashed?

Try pinging the server. Using one of the other computers, go to the "start" menu, select "RUN". type "command" in the resulting box. Pinging the server is akin to saying to that server "are you listening"?

Type PING and the name of the server.

[SCREEN-SHOT PICTURE OF THE PING PROGRAM WIN-DOW]

Type "Exit" to clear this window.

If you get anything else, there may be a problem with that computer's network connection.

Troubleshooting Bad Network Cards

If the network card has failed, it will manifest itself as a "dead card". The card's lights will not light (look at the back of the server).

Look at the back of the server, there is a silver plate on the back with a "phone cord-y" looking connector. Are the lights lit?

There may be 2 or three lights. LINK or ACT should be lit, ACT may be flickering. Flickering is good, it means that traffic is hitting the card.

If there is no flickering, it means that there is a connection problem.

If the network card in a server has failed, there is no easy fix. You will need to communicate this to the users.

Try Restarting the server to see if this corrects the problem.

THE INEVITABLE "RESTARTING THE SERVER" SECTION

One of the most deceptively simple things about a server is the ability to restart it, and hope for the best.

True, restarting the server usually corrects the problem, but the problem often recurs, when the person who originally caused the crash decides once again to print that document, or to try to send that corrupted email again "to see if it works this time".

Restarting the network servers should not be done with a cavalier attitude, and should not be the "last step" in correcting the problem.

It is bad to reboot, if it can be avoided. In Windows NT, many of the services which control the server processes (programs) themselves can be restarted, without causing the entire machine to lurch out of sync, and to have a 5 minute outage of the entire server. While some servers do not have problems with prolonged outages, others like MAIL and ADMIN, have significant repercussions associated with rebooting, or prolonged outage.

The specific instructions for restarting each server are discussed later in this document

Typically though, you will want to make sure that

- There are no CDs or Disks in the CD or Diskette drives of the server which you are restarting (tapes make no difference)

- The UPS is not showing an error light

WHAT IF THE SERVER IS SIMPLY "OFF" and won't come back on?

The most common problem with computers lies in the power supply. If a computer repeatedly crashes, inexplicably, or has appeared to turn off, but won't come on, it is likely that the server has succumbed to a power supply failure.

First, check the server's power cord (on the back), to verify it's electrical connection. If a power supply fails, there is little that you can do about this.

CHECKING THE UPS

Check the UPS and the power strips. The UPS is a battery backup device for a server. Ideally, in the event of a power outage, the UPS would take over the power requirements of the server, indicate to the server that it needs to shut down, and power the server while it does (automatically) an orderly shutdown.

If the UPS fails, there may be a service outage, as it is connected IN LINE with the equipment's power strip.

If a UPS fails, call the parish accounting office to get in touch with Coastal Business Machines. Coastal does on-site repair of the UPS equipment. DO NOT ATTEMPT TO DISCONNECT OR MOVE THE UPSs, as they can easily cause fires or electrocute you, and are very heavy and can injure you.

CHECKING THE POWER STRIP

Did something fall on the strip? Did the strip fail?
Look for the switch to make certain that the strip is still plugged in, and getting power.
If the powerstrip fails, you will need to replace it.
CHECKING THE ELECTRICAL SERVICE

If there was a storm, especially. All of the power service in the system room is outside in the ante-room to the system room. Check the circuits in the circuit breaker on the right hand side.

Facility Management, The Plant Manager, and The Electrical Contractor should be contacted with any electrical anomaly.

WHAT IF THE SERVER WON'T RESTART BECAUSE THE HARD DISK HAS FAILED?

If a hard disk or RAM fails, it is likely that the server will appear "frozen in time". There may be gibberish on the screen, or nothing on the screen.

If you restart the server, and it indicates a "disk failure", or "ram failure". There is a serious hardware problem. You should not go any further, because data on these drives may be recoverable, and going forward may defeat this.

A Word on ISSUES WHICH YOU SHOULD NOT TRY TO RESOLVE

What if a server has failed in one of the above ways, you may not be able to repair it.

It is key to note that the servers SHOULD NOT be removed from the racks. These server cases are much heavier and bulkier than normal PC cases. If something fails that seems to be hardware related, please do not attempt to remove a server from the rack. Doing so can cause you significant injury, and may do damage to that server, or to other equipment in the rack.

LIGHTNING STRATEGY

Lightning likes our network. It seems that, especially in spring and summer, we have numerous outages due to lightning.

Key systems which are almost always affected:

- The Internet Router

- The Telephone System

- The Alarm System

TELEPHONES AND LIGHTNING

The telephone system seems to have a problem every time that there is lightning or an electrical storm. There is usually nothing to be done about this, short of contacting Interconnect services.

THE ALARM SYSTEM AND LIGHTNING

The alarm system usually makes it through lightning storms, but occasionally the burglar alarm will have some sort of failure.

The key here is to contact [The Alarm System Contractor] as soon as possible. It may also be advisable to go to the parish after a lightning storm after hours in order to determine any damage.

DAMAGE TO THE ALARM SYSTEM IS REPORTED IN ONE OR MORE OF THREE WAYS:

- ERROR INDICATORS ON THE READ-OUT ALARM PANEL IN THE BOILER ROOM

- ERROR BEEPS ON THE PANELS

- CALLS FROM ALARM MONITORING TO US ABOUT PROBLEMS.

THE COMPUTER SYSTEM AND LIGHTNING

The computer system is fairly protected from lightning. The outlets are recent, and grounded using GFCI. The wires for the network are made of glass, not metal, and conduct no electricity.

The only issues which have arisen in the past have to do with the routers and switches. Occasionally, the operation of the routers and switches will be taken out by storms.

Contact [The DSL provider], if you determine a problem with the router, or if the Internet is out after a storm. Also, Contact [The Network Equipment Contractor]. if a switch seems to be out.

MAKING DUE WITHOUT SERVICES

This document touches on the iceberg, as it were, on the multitude of potential problems which can occur with network servers. A comprehensive backup plan would not be complete without a section on making due without use of network services. Especially in the typical situation, where the network administrator is away for a week or so, the network server may be out for a few days.

Depending on the system, you may or may not have a significant problem. Look at the sections on each server for more details on what is out if it fails.

Fall-Back procedures:

TELEPHONE—
Real telephones can be connected through the patch panels. If the phone switch fails, call ISI from a non-switch "real" phone. Even if a person must be stationed at a telephone, The main number can be connected to a live operator with a phone, who can continue to conduct business.

Electronic Mail—
If the email fails, it is most likely on our end. The incoming email will be sent back to the recipients. As a backup plan, a new server with the mail server may need to be set up. Contact the Vendor listed on the server's service tag (on the back of the server).

If for some reason email communication is critical, it is also possible to setup free email accounts on
www.hotmail.com
www.yahoo.com

INTERNET ACCESS & ACCESS TO OUR WEBSITE
The Internet can be accessed by telephone dial-up connection

[LIST PROVIDOR NAME AND DIAL-UP LINE NUMBER]
[LIST NAME AND PASSWORD FOR ACCOUNT]

FAILURE OF NETWORK LOGINS

In the unlikely event of the failure of both the [PDC] and [BDC] servers simultaneously, there would be the inability to log in to your PC, because the network would not be able to validate your login name.

You would still, however, be able to log in to your own computer, but will receive an error message when you do so. Your network resources, like printers, would be unavailable to you.

If you are using windows 98, you will need to hit the "ESCAPE" key to bypass the network login.

PRINTING

If the network print server fails, you can connect the printers using a parallel cable to a computer, thus setting up "print stations".

Step 1: Procure a parallel cable from a computer supply store. Connect the printer to the computer.

Step 2: Add a printer, using the print wizard in, "START", "SET-TINGS", "PRINTERS", to add an HP LaserJet to port LPT1. You should be able to print.

APPENDIX C

"PDC" AND "BDC" SERVERS

It is a good idea to keep a description sheet about each of your servers. Here is a sample sheet that we keep attached to each server, in the event that it needs service and the network administrator is not available.

TITLE: SERVER SPECIFIC SPEC SHEETS AND BASIC TROUBLESHOOTING
VERSION INFORMATION: APRIL 2000
UPDATE RESPONSIBILITY: JOE LIBERTO/SYSTEM ADMINISTRATOR
PAGE 1 OF 1

SPECIFICS:
These servers are H-P tower type case computers which are mounted in the rack with the mail server.

NETBIOS NAME: ADMIN/ACADEMIC
IP ADDRESS (EXTERNAL): none
SUBNET MASK (EXTERNAL): none
IP ADDRESS (INTERNAL): 10.0.0.2/10.0.0.3
SUBNET MASK (INTERNAL): 255.255.255.0

WHERE IS IT?

Facing the rack, it is on the left hand side, top PC.
ADMIN is labeled "ADMIN"
ACADEMIC is labeled "ACADEMIC"

WHAT OPERATING SYSTEMS DO THEY RUN?

Windows® NT™ server 4.0

When are they down?

If the server is not responding to mouse or keyboard input, or not responding correctly

If windows are redrawing slowly, or won't move on the screen

If it will not respond to being Pinged, or if the LINK or ACTIVITY lights are not lit—it may be difficult to see these lights, due to the placement of this server.

What are the ramifications of these servers failing?

If this primary server (ADMIN) fails, no personal drive data will be available.

If the backup server (ACADEMIC) fails, no accounting or shared parish data will be available. No PDS data will be available

If BOTH fail, in addition to the above problems for each server, no logins will be processed, no SJS network authentication can occur, no printing will occur.

REBOOTING THESE SERVERS

If a problem has occurred with either server, the best course of action is to reboot them.

Using the KDM switch, select the errant server. It may have a blue screen. If this is the case, simply use the power button on that server to recycle power to that computer.

If the screen looks normal, log into that computer using the Administrative account.

To Restart:

Log in, and Click the Shutdown option from the START MENU

Click on the button marked "RESTART"
Click "OK" and wait.

Eventually, after the reboot, the computer will show the familiar "Press CTRL+ALT+DEL to LOG IN" screen. Do so, and log in as the ADMINISTRATOR.

APPENDIX D

PRINTER RELATED ISSUES FROM THE EMERGENCY PLAN

This is a good printer trouble-shooting template.

<u>PRINTING RELATED ISSUES</u>

<u>PRINTER DIALOGUE BOXES</u>

The most common problem with the print server is that an error message (or series of error messages) will appear on the screen of the server. These error messages will be in the form of "cannot print to NSOFFC on port NE3" or some similar wording. Clicking OK may dismiss the message, or may just bring up another similar or duplicated message. The most common fix is to stop and restart the printer spooler service on the server.

See below for more info on this.

<u>DOES THE PRINTER HAVE PAPER?</u>
Usually, printers pass error messages from the print server on to the client PC, so the client PC will get the "out of paper" message. Occasionally, though, the print server will hold the error message and it will appear on the screen of the print server, rather than on the screen of the User's PC. This is most common with older LaserJet and DeskJet

printers which are connected to print sharing devices. Check to see that there is paper in the printer, and that if it is a laser printer, that the printer's online light is lit.

DESKJET AND LINE PRINTER ISSUES

Sometimes, DeskJet printers and line printers will fail. Usually, it is not the printer itself, it is a failure in the external device. This particular problem manifests itself especially in the older "Linksys" brand print sharing devices (Check for the Linksys Logo on the print-sharing box). Print sharing devices on one end connect to a printer, and on the other end connect to the network jack. They are small plastic boxes, located in close proximity to the printer. They are white in color, usually have three wires coming out of them, one leading to the wall jack, one to the power outlet, and one larger cable to the printer itself.

You may need to reset the box by unplugging it from the wall, to correct any problems with the print sharing device.

STOPPING AND RESTARTING THE PRINTER SPOOLER SERVICE ON THE PRINT SERVER
(See the Print Server diagnostic sheet for introductory Information)

1. Click on the "Start Menu"

2. click on "SETTINGS" and "CONTROL PANEL"

    the control panel will appear on the screen

3. click on the "SERVICES" icon

    the services control panel will appear on the screen

4. scroll down the list until you see the SPOOLER service.

5. click once on SPOOLER

6. Click the "STOP" button and wait.

7. when you are given control back to the window, click the "START" button to restart the spooler service.

8. Click OK to close the window.

9. Click the X button in the top right corner to close the Control Panel window.

Appendix E

SAMPLE BACKUP INFORMATION SHEET

I wrote this document to instruct a person who was basically non-technical on how to perform the data-backup tape rotation. I did this because the tapes must be rotated daily. In the event of my absence, either through unexpected illness or planned vacation, the tapes will continue to be changed. There is a particular person whose responsibility it is to perform this duty in my stead. Notice the structure of this document, and the types of information described therein. This document was stored on our document server, which does not allow for graphical information. This is why the document is in a typewriter typeface, and why there are no graphics involved in the explanation.

TITLE: BACKUP ROTATION SCHEDULE INSTRUCTIONS
VERSION INFORMATION: MARCH 2001
UPDATE RESPONSIBILITY: JOE LIBERTO/SYSTEM ADMIN-
ISTRATOR
PAGE 1 OF 3

1. HOW TO CHANGE THE TAPES

2. HOW TO MARK THE CALENDAR AFTER YOU
   CHANGED THE TAPE

APPLICATION
This document will assist someone who is to carry out the daily task of
changing the backup tapes

WHY IS THIS IMPORTANT?
The chief product of the computer network is the data produced and
stored on the network servers.
In the event of a server crash, your only reliable means of data restoration is from the data backup tapes.
Daily changing of the tapes ensures a full 15 (x2) tape redundancy. 15
day (x2) redundancy is important because errors
or loss of data may not be immediately apparent.
Why (x2)? There are two separate tape backup devices. The tapes are
cycled independently, and perform 2 identical backups. One backup
machine is isolated for the purposes of backup, to increase the odds of
data protection.

HOW DOES THE BACKUP CYCLE WORK?
There are 15 tapes, numbered on one side 1 through 15, and on the
other side 16 through 30. There is no tape 31.
So, the label on the first tape in the set (1) looks like this:

```
------------------------
```
1–15 (1) 8/01
```
------------------------
```

1–15 <—the tape number
(1) <—indicates that this is the set 1 tape drive (DDS1 type tape)
8/01 <—indicates that this tape was put into service in august of 2001
and the label on the first tape in the set (3) looks like this:

```
------------------------
```
1–15 (3) 8/01
```
------------------------
```

1–15 <—the tape number
(3) <—indicates that this is the set 2 tape drive (DDS3 type tape)
8/01 <—indicates that this tape was put into service in august of 2001

TITLE: BACKUP ROTATION SCHEDULE INSTRUCTIONS
VERSION INFORMATION: MARCH 2001
UPDATE RESPONSIBILITY: JOE LIBERTO/SYSTEM ADMIN-ISTRATOR
PAGE 2 OF 3

<u>WHAT DO I DO?</u>
In the morning, go to the server room. You will need a MP master key to get into the server room.
<u>Step 1</u>: Look at your watch or a calendar for the date. If today is the 31st, do not do anything.
<u>Step 2</u>: Go to the DDS-1 box. The DDS-1 box is a Grey/manila plastic box on the rack next to the tape backup server. There are two boxes, one is marked "<u>DDS-1 tapes</u>", the other is marked "<u>DDS-3 tapes</u>". Open the box, take out the tape with today's date on it.
For example: If today is the third, take the tape labeled <u>3–18 </u>out of the box.
<u>Step 3</u>: Go to the server labeled "8TRACK". 8TRACK IS A White colored DELL POWEREDGE computer.
ON THE FRONT OF THE DELL, IS A TAPE DRIVE, located directly below the CD-ROM Drive.

IS A YELLOW LIGHT BLINKING ON THE DRIVE? IF SO, GO TO SECTION 2

<u>Step 4</u>: FIRMLY Press the only button on the tape drive. The button is located below the hole for the tape. Pressing this button will eject the tape from the drive. Ejecting the tape takes several minutes. <u>Please be patient</u>. After the tape drive automatically rewinds the tape, the tape will pop out of the drive.

Step 5: Take the tape from the drive, and put it into the box marked with the same label as the tape. If the tape label says **1–16**, put the tape in the box marked

1–16 (1) 8/01

Step 6: Take today's tape out of the plastic box, place the box on it's side back in the plastic tape storage box marked "DDS-1 TAPES".

Step 7: Push the tape FIRMLY into the Slot for the tape (where the old tape ejected from) until the tape is pulled into the drive.

DOES THE NEW TAPE KEEP POPPING OUT OF THE DRIVE? IF SO, GO TO SECTION 3.

Now, you're not finished.

Remember, this is a dual backup system, you need to change the other tape.

Step 8: Open the Grey box marked "DDS-3 Tapes". Take today's tape from the box.

Remove the tape cassette from the clear plastic cover, and return the plastic cover to the Grey box.

Step 9: The other tape backup unit is attached to a computer located in a rack, at about waist height if you are facing the rack. The rack is opposite to the first tape backup computer. The computer is one of two identical machines, and is on the left hand side. The computer is light Grey, and is labeled "ADMIN", and the tape drive is a small box sitting on top of the computer.

Step 10: Press the button below the hole for the tape **firmly** to eject the old tape. BE PATIENT, this will take a few minutes.

Step 11: The old tape will pop out of the drive, put the tape back in the clear plastic covering, and return to the box marked "DDS-3 Tapes".

Step 12: Slide Today's tape into the slot, and press firmly until the tape is pulled into the drive.

LAST STEP: On the wall next to the door is a calendar marked "Backup Calendar". The backup calendar helps you to keep track of which tape was used on which day. This is important for the purposes of restoring data from tapes used previously.
MARK THE CALENDAR ON THE WALL MARKED "BACKUP CALENDAR" IN TODAY'S BLOCK WITH THE TAPE NUMBER THAT YOU USED.
FOR INSTANCE:

Wednesday

3
3–18 <—you would write "3–18", as you used the "3–18" tape today
----

NOTE: IF YOU FORGOT TO CHANGE A TAPE YESTERDAY, MARK YESTERDAY'S BLOCK WITH THE PREVIOUS DAY'S NUMBER, TO INDICATE THAT YESTERDAY'S BACKUP OVERWROTE THE PREVIOUS DAY'S DATA ON THE TAPE.
Example: Today is Friday the 5th, you forgot to change the tape on Thursday the 4th. Write "3–18" in Thursday the 4th's block, to indicate that the backup used the "3–18" tape. Because you failed to change the tape, the 3–18 tape was still in the drive on Thursday when the backup automatically occurred, so the 3–18 tape was used twice.
BACKUP CALENDAR

Wednesday Thursday Friday (today)
--------- --------- ---------
3   **4**   **5**

3–18   3–18   <—you would write this in on Friday)
--------- --------- ---------

END OF INSTRUCTIONS

TITLE: BACKUP ROTATION SCHEDULE INSTRUCTIONS
VERSION INFORMATION: MARCH 2001
UPDATE RESPONSIBILITY: JOE LIBERTO/SYSTEM ADMIN-
ISTRATOR
PAGE 3 OF 3

PROBLEMS WHICH MAY OCCUR
SECTION 2: IS AN ORANGE LIGHT ON ONE OF THE TAPE DRIVES BLINKING?

This means that the drive heads are dirty. In the Grey tape storage boxes are head cleaning cassettes. To use, eject the tape from the culprit drive by pressing the button on the drive, insert the cleaning cassette (labeled "Cleaning cassette"), wait about a minute. The cleaning cassette should pop out of the drive automatically. The light should stop blinking, and the heads are clean.

SECTION 3: DOES THE TAPE KEEP POPPING OUT OF THE DRIVE ON the "8TRACK" computer?

8Track's tape unit is older than the one on ADMIN. 8TRACK uses a DDS-1 type tape, and will not work with Admin's DDS-3 tapes. The DDS-3 tapes have a "DDS3" logo on them, and will eject from 8Track's drive, automatically.
Verify that the tape has a "DDS" logo on it, and was from the tape storage box labeled "DDS-1 TAPES"

0-595-26596-0

www.ingramcontent.com/pod-product-compliance
Lightning Source LLC
Chambersburg PA
CBHW051223050326
40689CB00007B/779